W9-AQR-793

A lucidly written and thorough approach to comprehending pollution issues, *Understanding Environmental Pollution* systematically addresses the spectrum of these issues, from global stratospheric ozone depletion to personal pollution in the home. Dr. Hill identifies pollutants and their sources, addresses the risks to humans and the environment, and discusses what is being done to reduce emissions. Throughout, she applies four tests to the relevant issue: What are the pollutants of concern; why are they of concern; to what sources do humans risk exposure; how are emissions of the pollutants being reduced?

Although it focuses on the United States, this text considers the global implications of local pollution and stresses both individual and corporate responsibility. The book provides basic concepts in toxicology and risk assessment along with dissenting opinions on major issues.

In presenting the facts and offering a variety of opinions, this book enables curious undergraduate students to evaluate environmental problems, realizing that there may not be a "right" answer. It is a perfect text for introducing students to pollution problems and to earth science.

UNDERSTANDING ENVIRONMENTAL POLLUTION

UNDERSTANDING ENVIRONMENTAL POLLUTION

MARQUITA K. HILL

University of Maine, Orono

CAMBRIDGE
UNIVERSITY PRESS

PUBLISHED BY THE PRESS SYNDICATE OF THE UNIVERSITY OF CAMBRIDGE
The Pitt Building, Trumpington Street, Cambridge CB2 1RP, United Kingdom

CAMBRIDGE UNIVERSITY PRESS
The Edinburgh Building, Cambridge CB2 2RU, United Kingdom
40 West 20th Street, New York, NY 10011-4211, USA
10 Stamford Road, Oakleigh, Melbourne 3166, Australia

First published 1997

Printed in the United States of America

Typeset in New Baskerville and Futura

Library of Congress Cataloging-in-Publication Data
Hill, Marquita K.
Understanding environmental pollution / Marquita K.
Hill.
   p.   cm.
Includes index.
ISBN 0-521-56210-4 (hc). – ISBN 0-521-56680-0 (pbk.)
1. Pollution.   2. Pollutants.   I. Title.
TD174.H55   1997   96–52929
363.73 – dc21   CIP

*A catalog record for this book is available from
the British Library*

ISBN 0 521 56210 4 hardback
ISBN 0 521 56680 0 paperback

Dedicated with affection and appreciation to

Clarence Owens of Jackson, Michigan,
humanist, naturalist, teacher

CONTENTS

PREFACE AND ACKNOWLEDGMENTS

Understanding Environmental Pollution grew out of a course first taught at the University of Maine in 1993 as part of a series of courses on technology and society. Not finding a text addressing pollution issues at a level suitable to a mixture of science and nonscience undergraduate students, I prepared class notes, which evolved into chapters and finally into the present book. The book's intent is to summarize the basics of a sometimes bewildering number of pollution issues in a systematic way. Because the students for whom the book was written come from a variety of backgrounds, the terminology is intended to be understandable to the educated nonscientist, while not being overly simplified for science and engineering students.

Initially, I had no intent to address social issues, but it quickly became clear that pollution cannot be dealt with, or even described, in the absence of people's understanding of and concerns about it. Even the laws that govern pollution change the way it is perceived. Thus, when a major controversy is associated with an issue, it is described along with the scientific uncertainty at its root. For example, some scientists believe the U.S. Environmental Protection Agency (EPA) exaggerates the dangers of radon in our homes, and some are not convinced that a significant amount of global warming will occur. Why? Readers are encouraged to go beyond a belief that there are always straightforward solutions to problems. Consider chlorofluorocarbons (CFCs,) which in the 1920s were seen as an answer to a problem. Looking for CFC alternatives in recent years raised problems presumably solved 70 years ago.

Although the analysis emphasizes U.S. pollution issues, international implications and comparisons are often made. Consider the recent finding that acid rain precursors, formed in the United States, are reaching Europe, or current concerns regarding the accumulation of polychlorinated organic compounds in the Arctic. Chapter 6 is devoted to three major global change problems that are associated with air pollution, acidic deposition, stratospheric ozone depletion, and global warming.

Thirty years ago, a question asked in Chapter 1, "What is pollution?" would have been answered differently than today. Pollution was obvious in the United States and, in many cases, it still is. But modern analytic tools detect increasingly tiny amounts of a chemical. So today we almost need an additional question: "At what point is so little of a chemical present that it ceases to be a pollutant?" Chapter 1 also examines limitations of the command-and-control approach to limiting pollutant emissions and provides a segue to pollution prevention, a concept introduced in Chapter 2 and emphasized thereafter. Basic concepts in toxicology are presented in Chapter 3 to explain why we care about pollutants, and Chapter 4 describes the use of risk assessment as a tool to evaluate environmental problems in the absence of absolute answers. Beginning in Chapter 5, four questions are asked for the pollutants examined: What are the pollutants of concern? Why are they of concern? What are their sources, especially sources of human exposure? What is being done to control and reduce emissions? Chapters 5 through 8 look at pollution from the viewpoint of the medium that is contaminated: air, water, or land. Subsequent chapters examine contamination of all these media with several of the pollutant categories that pose special concerns: metals, pesticides, and environmental estrogens. Because energy production and use are major pollution sources, a chapter is devoted to these topics. Finally, the practical issue of pollution at home is examined. The importance of personal actions is recognized throughout the book, and information is often provided on steps individuals can take within and beyond the home.

My appreciation goes to the following individuals, who reviewed one or more chapters of an earlier draft of this manuscript: At the University of Maine, Randall Alford, Professor, Applied Ecology and Environmental Science; Richard Hill, Professor Emeritus, Mechanical Engineering; Nicholas Houtman, Science Writer, Office of Public Affairs; Betty Ingraham, special assistant to the pilot plant, Department of Chemical Engineering; Victoria Justus, Director, Environmental Health and Safety; Stephen Kahl, Manager, Sawyer Research Center; Kenneth Mumme, Professor, Chemical Engineering; and Sharon Tisher, Adjunct Professor, Resource Economics and Policy. Others who reviewed chapters are Stephen Groves, former Director, Water Bureau, Maine Department of Environmental Protection; Stephen Huntley, Health Scientist, ChemRisk, Portland; John Selner, M.D., Denver, Colorado; Paul Shapero, M.D., Bangor, Maine; and I. Glenn Sipes, University of Arizona, Tucson. Special thanks go to two individuals who commented on the entire manuscript, Robert Graves, M.D., Orono, and Cynthia Hassler, graduate student at the University of Florida, Gainesville. Responsibility for errors obviously remains with the author. Thanks are also due to David Kraske, Professor Emeritus of Chemical Engineering, University of Maine, who made me mindful of the "fog index" mode of writing. Finally, I offer my love and gratitude to my husband, John C. Hassler, Professor of Chemical Engineering, who has given his support throughout this process while patiently maintaining my computer, its files, and its eccentricities on an ongoing basis.

ABBREVIATIONS AND ACRONYMS

ATSDR	Agency for Toxic Substances Disease Registry
CDC	U.S. Centers for Disease Control and Prevention
CFCs	Chlorofluorocarbons
CO	Carbon monoxide
CO_2	Carbon dioxide
CPSD	Consumer Product Safety Division
DDE	Dichlorodiphenyldichloroethene, a DDT degradation product
DDT	Dichlorodiphenyltrichloroethane
DES	Diethyl stilbestrol
dioxin	Refers to 2,3,7,8-TCDD or, sometimes, to the whole dioxin family
DOE	U.S. Department of Energy
EPA	U.S. Environmental Protection Agency
ETS	Environmental tobacco smoke
FIFRA	Federal Insecticide Fungicide and Rodenticide Act
FDA	U.S. Food and Drug Administration
FFDCA	Federal Food Drug and Cosmetic Act
HAPs	Hazardous air pollutants, also referred to as toxic air pollutants
HHW	Household hazardous waste
IPM	Integrated pest management
IR	Infrared
MCL	Maximum contaminant level
MEI	Maximally exposed individual
μg/dl	Micrograms per deciliter
MSW	Municipal solid waste
NAS	National Academy of Sciences
NASA	National Aeronautics and Space Administration
NOAA	National Oceanic and Atmospheric Administration
NOAEL	No observed adverse effect level
NRC	National Research Council, an arm of the NAS
NTP	National Toxicology Program

PAHs   Polycyclic aromatic hydrocarbons
PCBs   Polychlorinated biphenyls
pCi/l   Picocuries per liter
P^2 pollution prevention
ppb   parts per billion
ppm   parts per million
ppt   parts per trillion
RfD   Reference dose
TCDD   2,3,7,8-tetrachlorodibenzo-*p*-dioxin, also called "dioxin"
TUR   Toxics use reduction
UN   United Nations
UNEP   UN Environmental Program
USDA   U.S. Department of Agriculture
UV   Ultraviolet
VOCs   Volatile organic compounds

1

UNDERSTANDING POLLUTION

WHAT IS A POLLUTANT?

The United States Environmental Protection Agency (EPA) reports that Americans spend $140 billion a year to control and clean up pollution. By the year 2000, compliance with environmental regulations will cost an estimated $160 billion a year, 2.8% of the gross national product. How is this money spent? What is a pollutant? The United States EPA defines a *pollutant* as any substance introduced into the environment that adversely affects the usefulness of a resource. In the early years of the American environmental movement, pollution was easy to define and its adverse effects were easy to see. Some rivers were visibly polluted or had an unpleasant odor. The infamous Cuyahoga River in Ohio became so polluted it twice caught on fire from oil floating on its surface. Air pollution from automobiles and unregulated industrial facilities was obvious. In industrial cities, soot often drifted onto buildings and clothing and into homes. Severe air pollution episodes increased hospital admissions and killed sensitive people. See Figure 1.1 for a view of New York City blanketed by heavy smog in 1963. Fish and birds were killed by unregulated pesticide use. Trash was discarded in open dumps and burned.

Compared to that of 25 years ago, American air quality is much improved, although pollution is sometimes still obvious – traffic fumes in city traffic, ozone levels that irritate the eyes and lungs, or sulfate hazes that obscure the view. However, even when pollution is not visible and no adverse health effects are noticed, air may still contain pollutants at levels that provoke concern. Water quality has also improved over that of the 1960s and pollution is less obvious; however, although a river, lake, or bay may not be visibly polluted, may not have an unpleasant smell, and may have plentiful fish, the water or its sediments may contain synthetic chemicals. The word *synthetic* is deliberately used because there are always thousands of natural chemicals present. If the level of *anthropogenic* – human-generated – chemicals is high enough in a

1

FIGURE 1.1. New York City blanketed by smog, 1963. *Source*: AP/Wide World Photos. Reproduced with permission.

water body to affect bird reproduction adversely or to cause tumors in fish, all would agree that the water is polluted whether or not the pollution is visible. However, if only trace or low levels are present, and if there is no observed adverse effect in humans, animals, or plants, characterizing water as polluted is more controversial. Many still consider such water polluted, arguing that there may be adverse long-term consequences of even very low concentrations. Others would counter that very low levels of contamination do not have adverse effects and that we need to devote limited resources to high-risk problems.

Consider a situation where there are many synthetic organic chemicals present in a water or soil sample. Assume that each is present only at a tiny level, a level that few would consider a risk. Some are concerned about possible additive effects; that is, if a number of the chemicals exert biological effects by similar means, then, if the levels of all the chemicals are added together, the total can cause adverse effects. Worse, the effect may be synergistic; that is, one chemical may magnify the effect of another. Others counter that there are also thousands of natural chemicals present in water, soil, and food, many of which are similar, or identical, to anthropogenic chemicals. They further point out that an animal's body cannot tell whether a chemical comes from a natural source or from a human source. One qualification to this last statement is that humans, animals, and most microorganisms degrade certain polychlorinated chemicals extremely slowly.

Examples of polychlorinated chemicals are the pesticide dichlorodiphenyltrichloroethane (DDT) and many of the family of chemicals called polychlorinated biphenyls (PCBs). Literally thousands of chlorinated chemicals are produced naturally. However, levels of natural polychlorinated chemicals are very low. In the United States and other Western countries, many polychlorinated chemicals such as DDT and PCBs have been banned, with the result that environmental levels are now lower. However, as will be discussed later, there are hot spots still containing high levels of these pollutants.

Keeping in mind that many chemicals generated by human activities are also produced naturally, consider the following questions. What if society wanted to eliminate anthropogenic emissions of a certain chemical totally? Would it be possible to tell whether emissions are eliminated if natural sources still exist? This is not a theoretical question. Some manufacturing concerns have complained that a particular chemical is more concentrated in river water entering their facilities than is allowed by law in effluent water leaving them. A pollutant by definition produces undesired environmental effects. Thus, natural chemicals can also be pollutants. Consider the huge quantities of ash, chlorine, sulfur dioxide, and other chemicals released during volcanic eruptions. Or consider the radioactive chemical radon, which is ubiquitous in the environment from natural sources. Radon is associated with human lung cancer, and EPA ranks it second only to environmental tobacco smoke (ETS) as an environmental health risk. Radon levels in outside air are low enough that it is not considered a pollutant. However, when a home is built, the enclosed structure allows radon to become concentrated within it. Radon then becomes a pollutant. Consider the chemical arsenic, made famous many years ago in the play *Arsenic and Old Lace*. In some locales, enough arsenic is naturally present in drinking water to increase the cancer risk of persons drinking it by as much as 1%. There are many other examples of natural pollutants.

Scientists can provide guidance as to when a specific chemical is a pollutant. But they cannot say with certainty when the level of that chemical is so

TABLE 1.1 Terms used to describe pollutant
concentration

ppm = parts per million *a, b*
ppb = parts per billion   (one thousand times smaller than ppm)
ppt = parts per trillion   (one million times smaller than ppm)
ppq = parts per quadrillion (one billion times smaller than ppm)

a These terms refer to parts by weight in soil, water, or food or
   parts per volume in air, and the following provides examples of
   their meaning: 1 ppm = 1 pound contaminant in 500 tons (500
   tons = 1 million pounds); 1 ppb = 1 pound of contaminant in
   500,000 tons; 1 ppt = 1 pound of contaminant in 500,000,000
   tons; 1 ppq = 1 pound of contaminant in 500,000,000,000 tons.
b To gain perspective, compare ppm, ppb, etc., to periods of time:
   1 ppm is equivalent to 1 second in 11.6 days; 1 ppb is equivalent
   to 1 second in 32 years; 1 ppt is equivalent to 1 second in 32,000
   years; 1 ppq is equivalent to 1 second in 32,000,000 years.

low that there is no possibility that it will have adverse effects. In the end, society – each of us – must decide how strictly to define and regulate pollutants.
See Table 1.1 for a description of terms used to describe concentrations of environmental pollutants.

DISCUSSION QUESTIONS

Consider a pesticide that causes cancer in animals at high doses and
the question of whether it is a pollutant when found in trace (barely
detectable) amounts in food. Background information: The Delaney
Clause of the 1958 Federal Food Drug and Cosmetic Act (FFDCA) prohibited the addition to food of any amount of any chemical known to
cause cancer in people or animals. In 1958, when the law was passed,
few cancer-causing chemicals were known and the means by which they
caused cancer was not understood. This law also made no distinction
between potent carcinogens and very weak carcinogens. In 1958 analytical chemists could detect chemicals only at high concentrations,
one part per thousand or, sometimes, one part per million (ppm).
Thus, even a known carcinogen could escape regulation if present in
food below detection levels. In the 1990s, analytical chemists routinely
detect chemicals at parts per billion (ppb) or parts per trillion (ppt)

levels. Also, in the 1990s, hundreds of chemicals have been identified that when fed to animals in high doses induce cancer in those animals. A problem arises in finding a reliable method to relate cancer caused in animals exposed to high doses of a chemical to people exposed to low – sometimes extremely low – doses of that chemical. By the 1990s, scientists increasingly questioned the wisdom of the Delaney Clause. Even the U.S. EPA tried to ignore the letter of the law by allowing the use of carcinogenic pesticides in cases where it believed they posed a negligible risk. EPA was challenged in court for allowing this practice and lost. The court stated that the Delaney Clause dictating zero risk must be enforced until Congress chose to change that law. This meant EPA had to ban any pesticide that causes cancer in animals, no matter how weak the carcinogen or how tiny the amounts of pesticide residue. For the following questions, limit yourself to discussing food residues, not other problems associated with pesticide use.

(a) In your opinion, is any amount of a carcinogenic pesticide in food a pollutant?

(b) Pesticide residues on produce are measured in the field, but residues usually decrease after harvesting and during transportation and storage. Also, a large percentage of the residues are found on peelings or outside leaves, which can be discarded. Other residues are removed during washing or cooking. Does knowing this affect your opinion as to whether any amount detected on food is a pollutant?

(c) A 1993 National Academy of Sciences (NAS) report observed that children, unless evidence exists to the contrary, should be assumed to be more sensitive to pesticide residues than adults. Children are also more highly exposed because, pound for pound, they eat more than adults. Chemical risk assessments are now required to take children into account specifically. Does knowing

---

BOX 1.1

A different way of regulating a pollutant in food. Aflatoxin B_1 is a potent human carcinogen that is produced naturally by mold growing on peanuts and grains. If all foods containing a detectable level of aflatoxin B_1 were banned, we would eliminate a significant portion of our food supply. However, as natural rather than deliberate chemical additives, aflatoxins were not subject to the Delaney Clause. Instead, aflatoxin B_1 is assessed in terms of its *action level*: Only if a grain contains aflatoxins at 20 ppb or greater is it removed from the market.

BOX 1.2

What environmental health is not. To appreciate the meaning of a clean environment, consider the opposite, the severe pollution and environmental degradation found in many poor countries around the world. One of the grimmest pictures is that of Africa, as described in the December 1994 issue of *Environmental Health Perspectives.* Of 450 million people in sub-Saharan Africa, about two-thirds do not have safe drinking water. Rural dwellers may walk hours a day to obtain water, even water that may be contaminated with sewage, silt from soil erosion, fertilizers, pesticides, mine tailings, or industrial waste.

With a growth rate of 3% a year, Africa has the most rapidly increasing population in the world. Food production in Africa is not sustainable. More and more land has been taken over for agriculture and 70% of Africa's forests have been cut for farmland and firewood. Ever more intensive use of farmland has led to damaged topsoil and lowered agricultural productivity. Intensive agriculture has also contributed to desert formation – desertification – as has overgrazing by cattle and goats. Agricultural workers seldom know how to use pesticides safely, and millions of poisonings occur each year. Women's and children's health is compromised by direct exposure to smoke as they cook over open fires in small dwellings using fuels such as coal, wood, charcoal, crop residues, dung, and dried grass. These fuels are also burned for heating and light.

As population has increased and rural environments have become degraded, people have moved into increasingly crowded cities. There, most have unclean drinking water and very few have sewage disposal. Garbage collection is rare. Not surprisingly, infectious diseases are rampant. The situation is worsened by uncontrolled industrialization and mining. Factories are built within or near residential areas. With the exception of a minority that hold themselves to high standards whatever their location in the world, most companies make little or no effort to control the pollutants they produce. With few occupational health and safety standards, factory and mine workers are exposed to toxic substances and other dangerous working conditions. Petroleum companies have polluted water so badly in some locales that fish populations have been destroyed.

Compared to this devastation, American problems seem almost benign. Remember, though, that Africa's degraded environment was not always so. Africa's earlier traditional agriculture and stable population were sustained over many generations. European colonization led to the development of large plantations and the use of agriculture for cash and trade. At the same time, the population burgeoned. An environmental tragedy has been the result. The message for us is that without legislation and controls, an environment will degrade. Unless highly significant changes are made, Africa's future is bleak.

this affect your opinion as to whether any amount of a carcinogenic pesticide detected in food is a pollutant?

(d) A 1995 NAS report noted that the great majority of natural and synthetic chemicals found in food were unlikely to pose an appreciable cancer risk, even those with known carcinogenic potential. Of the report, one author commented, "Whether a chemical is man made or God made, it is still a chemical and the body handles it in similar ways." Does knowing this affect your opinion as to whether any amount of a carcinogenic pesticide in food is a pollutant?

In 1996, almost 40 years after the Delaney Clause was enacted, Congress replaced it with a new standard: "reasonable certainty" that no harm will result from exposure to a pesticide residue remaining on food. Environmentalists accepted this change because the new law, the Food Quality Protection Act, provided new protection for consumers. Now, before setting a tolerance (a tolerance is the residue that is legally allowed to remain on a food) level for a pesticide, EPA must consider not just cancer, but a broad range of health effects, including the special sensitivity of children.

CHEMICAL AND POLLUTANT CATEGORIES

Chemical categories

Chemicals pollutants are more easily understood if one first understands the chemicals of which they are composed. There are two major categories of chemicals, the organic chemicals and the inorganic (mineral) chemicals; organometallic chemicals can be considered a third category.

Organic chemicals. The word *organic* is commonly used to refer to what is "natural," as in "organic agriculture." However, to say that a chemical is organic simply means that it contains at least one carbon atom and, typically, more. Some organic chemicals, the hydrocarbons, contain only the elements hydrogen and carbon. Others often also contain additional elements, such as oxygen, nitrogen, and sulfur. Organic chemicals are synthesized naturally by microorganisms, plants, and animals. An example of a common organic chemical produced in nature is sucrose, or common table sugar, which is derived from sugarcane or sugar beets. Another is acetic acid, the chemical that gives vinegar its bite. Sucrose, acetic acid, and a great many other organic chemicals can also be synthesized by humans. Most of the chemicals that we call synthetic are only partially synthetic; chemists start with the natural organic chemicals found in petroleum, coal, or wood and modify these – sometimes greatly – to make plastics, drugs, pesticides, and other chemicals. The organic chemicals made in living organisms are called biochemicals. Some well known biochemicals are proteins, fats, and carbohydrates.

Inorganic chemicals. This second chemical category does not typically contain carbon. Examples of inorganic chemicals are sodium hydroxide (caustic soda or lye), sodium chloride (table salt), household ammonia, and the chlorine-containing chemicals found in household bleaches and swimming pool chemicals. Rocks, sand, metals, and metal salts are made up of inorganic chemicals. A few inorganic chemicals do contain carbon. Among these are washing soda (sodium carbonate), baking soda (sodium bicarbonate), and elemental carbon (carbon black and diamond). Living creatures also synthesize substances that are largely inorganic, such as bone, horn, and shell. About 98% of tooth enamel is mineral or inorganic. Some inorganic chemicals, including many of the metals in our diet, are essential nutrients, necessary to the functioning of certain proteins and other biochemicals.

Organometallic chemicals. The organometallic chemicals contain both an organic and an inorganic component. An example is methylmercury, a common water pollutant. The methyl in this chemical is organic and the mercury is inorganic. As is the case with organic chemicals and inorganic chemicals, many organometallic chemicals are naturally produced in plants and animals, for example, the cobalt-containing vitamin B_{12}. Another example is the blood protein hemoglobin; most of the hemoglobin molecule is organic, but it also contains the metal iron.

Pollutant categories

The categories considered here are fairly logical, but others can be devised. Indeed, the air, water, and solid waste pollutants mentioned in later chapters have categories that are largely defined by law.

Organic pollutants. An organic chemical pollutant that is frequently mentioned in this text is PCBs. Most pesticide pollutants in the environment are also organic chemicals, for example, the once heavily used pesticide DDT and the carbamate pesticides. The organophosphate pesticides are analogous to organometallic chemicals in that they have an organic component plus an inorganic component, the phosphate (also see Table 1.2). Oil is another example of a substance largely composed of organic chemicals. It becomes a pollutant when spilled by accident. On a smaller, but ongoing scale, oil pollution also results when do-it-yourselfers change the oil in their cars and trucks and when motor boats and other recreational vehicles leak. Polluting oil is also

TABLE 1.2 Pollutant categories

CATEGORY AND EXAMPLES	ORGANIC OR INORGANIC?
Organic PCBs, oil, many pesticides	
Inorganic Salt, nitrate, metals	Salt and nitrate are inorganic, but metals may exist as organometallic pollutants (e.g., methylmercury).
Acid Sulfuric, nitric, hydrochloric	Examples shown are inorganic acids. Organic acid pollutants are produced in smaller amounts.
Physical Soil, trash	Both soil and trash have a variable composition of inorganic and organic components.
Radiological Radon, radium, uranium	The examples shown are elements and inorganic.
Biological Pathogenic microorganisms	Living organisms are largely organic, but do contain inorganic components.

found on streets, highways, parking lots, and construction sites. From these sites it can run off into lakes and streams or percolate into groundwater.

Inorganic pollutants. A common category of inorganic pollutants is the synthetic plant nutrients found in fertilizers. Nitrate, a component of fertilizers, is one example. Found in well water at a level above the U.S. EPA standard, nitrate is treated as immediately dangerous to life and health. The nitrogen and sulfur acids found in acid rain are also inorganic pollutants; salts are another example, including sodium chloride, common table salt. Rain or snow melt runoff from salt piles or roads treated with salt may contaminate well water to an extent that it becomes undrinkable. Metals are a major subcategory of inorganic pollutants. Metal emissions often are produced by industrial operations, but naturally occurring metals can also present problems. Arsenic is a natural component of bedrock in areas with granite and is sometimes found in drinking water at levels high enough to raise concern for human health. Mercury is found naturally in marine waters and some fresh waters, but it is also often the result of human activities. When electric power plants and other facilities burn coal, oil, or wastes that contain mercury, the volatile mercury escapes into the atmosphere and later settles onto land and water. In water, microorganisms can convert mercury into methylmercury – an organometallic chemical that is much more toxic than elemental mercury itself. Cadmium and lead are other metals released by these facilities.

Acid pollutants. The best known acid pollutants, sulfuric acid and nitric acid, have been mentioned as the inorganic pollutants found in acid rain. Acids are also found in runoff from coal mining and metal mining sites. Drainage from mines can severely acidify nearby water bodies. Another acidic inorganic pollutant is hydrogen chloride, a gas emitted by incinerators, which forms hydrochloric acid after reaction with moisture. There are many organic acids, but they are less common pollutants.

Physical pollutants. A physical pollutant is any solid material found in an inappropriate location. Soil, carried in rainwater runoff from agricultural fields, city streets, and construction sites, is an example of a physical pollutant. Trash is another example. Physical pollutants may have both inorganic and organic components. A different type of physical pollutant is the high-temperature water released from an electric power plant.

Radiological pollutants. The radiological pollutants are radioactive chemicals found naturally in rocks, water, and soil. In the waters of some midwestern states, radium may be naturally present in significant amounts, and the radioactive gas radon naturally occurs in significant amounts in some waters of New England states. Both radon and radium are elements and, as such, are inorganic chemicals. Anthropogenic sources of radioactive pollutants are the hazardous waste sites in locations formerly operated for military purposes.

Biological pollutants or pathogens. Pathogenic microorganisms are the biological pollutants ordinarily of most concern: infectious bacteria, viruses, and protozoa. Microorganisms are always naturally present in soil, water, air, and food, as well as on and within our bodies and those of all animals and plants. Both these microbes and those found in sewage or animal wastes can pose problems. The protozoan *Cryptosporidium* is a water pollutant that can cause illness and death when present in drinking water. Meat contaminated by certain bacteria can also cause serious illnesses and deaths. The broader term *biological pollutant* includes dead and living microorganisms and fragments of insects and other organisms that can contaminate air, water, or food.

Multiple pollutants. Rather than designating a category, the term *multiple pollutants* reflects the fact that living creatures are usually exposed to many pollu-

BOX 1.3

A preventable tragedy. Union Carbide, an American owned factory in Bhopal, India, manufactured the pesticides Temik and Sevin. The manufacturing process made use of the chemical methyl isocyanate (MIC). In addition to being an extremely toxic volatile liquid, MIC reacts violently with water. Despite this, the factory did not take adequate measures to prevent water from ever coming into contact with the MIC. On the night of December 2, 1984, as the people of Bhopal slept, water entered a storage tank that contained 50,000 gallons of MIC. The Indian government said that improper washing of lines caused the catastrophe. Union Carbide claimed that a disgruntled employee had deliberately introduced water into the tank. Whatever the cause, the resulting explosion released 40 tons of MIC and other chemicals over the city. Up to 2,500 people were killed overnight and the final death toll reached at least 6,000. Another 200,000 residents were injured, and, among these, many thousands suffered permanent injury. The situation was worsened by the fact that many people lived in crowded conditions close to the factory. Injured residents received little medical attention at the time of the accident, and compensation for their injuries was a long time in coming. Twelve years later, many remained ill. Americans reacted to the Bhopal tragedy with horror. They also discovered that the United States could experience a similar tragedy. A sister plant in Institute, West Virginia, that was also inadequately protected against the occurrence of a similar MIC accident was quickly redesigned after Bhopal.

Union Carbide was not unscathed. In 1984, it had almost 100,000 employees, but after Bhopal, it almost went out of business; by 1994, it was a much smaller corporation, employing only 13,000. In direct response to the Bhopal tragedy the United States Congress passed the Emergency Planning and Community Right-to-Know Act (EPCRA) in 1986. This law required industries and communities to prepare emergency plans to minimize harm in case of accidental chemical release. EPCRA gave Americans access to information on hazardous chemicals used in, stored in, or transported through their communities. An important part of the law, the Toxics Release Inventory (TRI), required businesses to report yearly on the emissions of 320 chemicals, including legal emissions. Examples of chemicals reported under the law are ammonia, hydrochloric acid, methanol, toluene, acetone, and lead and cadmium compounds. In 1994, the inventory was expanded to include hundreds of additional chemicals.

tants at one time. Some people are extremely concerned by these multiple exposures because they believe that many small exposures, added together, may cause a serious exposure. Others are more comfortable with multiple small exposures and point out that, over millions of years effective defenses have evolved in animals to protect them against small amounts of xenobiotics (foreign chemicals). Whether those xenobiotics are synthesized by humans or by the natural world usually matters little.

LIMITS TO POLLUTION CONTROL

Hazardous chemical legislation

Although environmental laws were passed in the United States before 1970, none tried to control pollution comprehensively. The 1970s saw a dramatic change. The first major law regulating pollution was the Clean Air Act (CAA) of 1970, which was followed by the Clean Water Act (CWA) in 1972 and the Safe Drinking Water Act (SDWA) in 1974. Legislators then turned their attention to land pollution and, in 1976, passed the Resource Conservation and Recovery Act (RCRA, pronounced "rick-rah") to control management and disposal of solid waste, including municipal and hazardous waste. After passing RCRA, many legislators felt they had effectively addressed the problem of pollution. However, the late 1970s revealed the existence of many abandoned hazardous waste sites around the country. Congress responded by passing the Comprehensive Environmental Response, Compensation, and Liability Act (CERCLA or Superfund) to clean up abandoned hazardous waste sites. Toxic chemical laws were also passed during the 1970s. The Federal Insecticide, Fungicide, and Rodenticide Act (FIFRA) was enacted in 1972; earlier forms of

---

BOX 1.4

Why does pollution occur? The Bhopal tragedy is an extreme event that should never have happened. Pollution in its everyday manifestations, however, cannot be completely eliminated: I am; therefore I pollute. This is true because no process is 100% efficient. Consider your own body, which cannot use 100% of the food ingested. Part of the food becomes waste material and part of its energy value becomes waste energy. The same is true of any process – waste material and waste energy are produced, and, unless the system is isolated from the environment, we cannot recover 100% of the heat and material waste: Some escapes as pollution. To understand this, consider several natural laws.

Conservation of energy. An important law of thermodynamics is that, in essence, energy can neither be created or destroyed. For example, in a power plant producing electricity by burning coal, none of the coal's energy value is lost as it is burned. However, according to the second law of thermodynamics, "water does not run uphill"; that is, the disorder (the entropy) of a system increases. In the case of an electric power plant, only about 40% of the coal's energy value is usable to run turbines to produce electricity. The other 60% is low-grade energy, which ordinarily becomes thermal pollution. Thermal pollution is found in steam released from power plant stacks, which heats the surrounding atmosphere, and in the plant's effluent, which increases the temperature of the water body to which it is released. A few plants make efforts to use the low-grade energy to heat buildings, greenhouses, or fish ponds. This action lowers, but does not eliminate, the loss of energy.

Conservation of mass. Another natural law may be stated as "Matter can be neither created or destroyed." In the example of the power plant burning coal, this law indicates that all the matter in the coal is conserved even though it is burned. However, chemical transformations do take place, as you can see from the following simplified reactions:

Coal component		Product formed
Carbon + atmospheric oxygen	$\rightarrow$	carbon dioxide
Hydrogen + atmospheric oxygen	$\rightarrow$	dihydrogen oxide (water!)
Sulfur + atmospheric oxygen	$\rightarrow$	sulfur dioxide
Metals + atmospheric oxygen	$\rightarrow$	metal oxides

As you might guess from these reactions, the total weight of products formed is greater than the weight of the coal burned. This is true because not only is the coal's matter conserved, so is that of the atmospheric oxygen with which it reacts, so the weight of the products formed equals the weight of the coal plus the weight of the oxygen with which it reacts.

FIFRA had not addressed pesticide toxicity. In 1976, the Toxic Substances Control Act (TSCA, pronounced "tosca") was passed to regulate chemicals that were not regulated under the laws already passed. Passing TSCA was an attempt to close the circle, to make sure that all chemical issues were addressed. For example, TSCA mandated controls for chemicals such as PCBs, which had not been covered by other legislation. A major mandate of TSCA was to develop an inventory of all commercial chemicals used in the United

States. Subsequently, any chemical not on the inventory would be subject to
scrutiny by EPA before it could be commercially used or imported. Toxic
chemicals became a more urgent issue after the 1984 tragedy in Bhopal, India
(see Box 1.3).

The limits of legislation

Laws controlling pollution are important and can be very effective. Their
shortcoming is that they only regulate pollution end-of-pipe: That is, pollutants are captured after they are formed. For example, an electric power plant
produces sulfur dioxide while burning coal. It then captures much of that sulfur dioxide from its stack gases to prevent it from entering the air. Or a metal
plating firm recovers chromium from its wastewater to prevent it from being
released into a river. The pollutants still exist, but in a captured form that can
be disposed of responsibly. End-of-pipe efforts significantly reduce air, water,
and land pollution. Industries and cities have built many wastewater treatment plants. Industry began to recover hazardous components from wastewater and air effluents and to incinerate or otherwise treat hazardous waste
before disposing of it. Irresponsible waste dumping was greatly reduced. To
replace old leaking landfills, municipalities and industries began to design
and build secure landfills for their wastes. Controls were placed on motor vehicle emissions.

   End-of-pipe control leaves us with the recovered wastes, which must be disposed of, potentially creating greater localized land pollution. In other instances, methods designed to remove volatile organic compounds (VOCs)
from water led to VOC release into the air instead. The costs of pollution control – recovering, treating, and disposing of the wastes recovered – continue
to rise. It is also costly to remediate sites at which wastes were dumped in a way
not considered acceptable today. Costs to municipalities of managing wastes
from our homes likewise continue to rise. The current $140 billion per year
that this country spends on controlling and cleaning up pollution may increase to the point that it will equal the amount spent on defense.

   Pollution control is never 100% effective in removing, treating, or disposing of pollutants. An industry or electric utility may remove the first 80% of a
pollutant from a waste stream in an efficient and cost-effective manner. However, removing the last 20% may cost 10 times more than removing the first
80%, and a small amount of the pollutant may remain. An incinerator may be
designed to destroy 99.99% of the hazardous materials that it burns. For pollutants of special concern, a more costly technologically sophisticated incinerator can destroy 99.9999%. But the incinerator cannot destroy that last tiny
amount. At the same time that we were finding limits to pollution control, we
were developing an ethic that says the only acceptable amount of a pollutant

is zero. This gives rise to the problem of the *vanishing zero* – even if that last bit of pollutant could be removed, another problem arises: Analytical chemists continue to devise ever more sensitive detection methods and the pollutant becomes detectable again. A chemical undetected and thus unregulated in one year could be detected and regulated the next.

One criticism of environmental laws and regulations made by some politicians and scientists is the assumption that one size fits all: A law not only tells industry to limit its pollutant releases, but also specifies exactly how the reduction must be accomplished. This is not satisfactory for many industrial facilities because no two are identical and frequently there is more than one way to reach the same end. Consider an example described in the March 29, 1993, *Wall Street Journal*: The AMOCO refinery of Yorktown, Virginia, was required to install a $41 million system using a specific type of technology to capture benzene emissions. However, AMOCO, working with EPA, found a different method that could capture five times as much pollutant for only $11 million. The law was inflexible and gave EPA no authority to modify requirements, so a greater amount of money was spent to capture less pollutant. In another example, regulations whose purpose was to ensure the safe handling of hazardous wastes had the unintended consequences described in a September 1995 *Scientific American* article: The automotive industry produces a sludge containing zinc as part of a process to protect steel from corrosion. In earlier years, the sludge was sent to a smelter to recover the zinc, which could then be reused in the process. However, in the 1980s, the zinc sludge was listed as a hazardous waste and the smelter would no longer accept it. Instead, the sludge became yet one more hazardous waste that had to be treated and disposed of in accordance with strict regulations. Clearly, these were unintended results. Most agree that problems like these arise because we are attempting to deal with one environmental problem at a time – bit by bit – instead of looking at issues holistically. Given the huge number of regulations that has resulted from our environmental laws, each specified in great detail, unintended and counterproductive results are inevitable. To correct problems of this nature, EPA has begun to develop regulations that, within the constraints of the laws passed by Congress, take into account all the emissions and waste streams issuing from a facility.

Politicians and other critics have recently dwelt on the problems resulting from environmental laws rather than the many successes. But where we would be without these laws? This question may be answered by examining the dismal state of the environment in Eastern Europe and the countries of the former Soviet Union, where strong environmental protection laws do not exist (also see Box 1.2). We obviously need environmental regulations, but most acknowledge that there is a limit to societal control of pollution. Consider that the United States generates more waste with each passing year and that it is in-

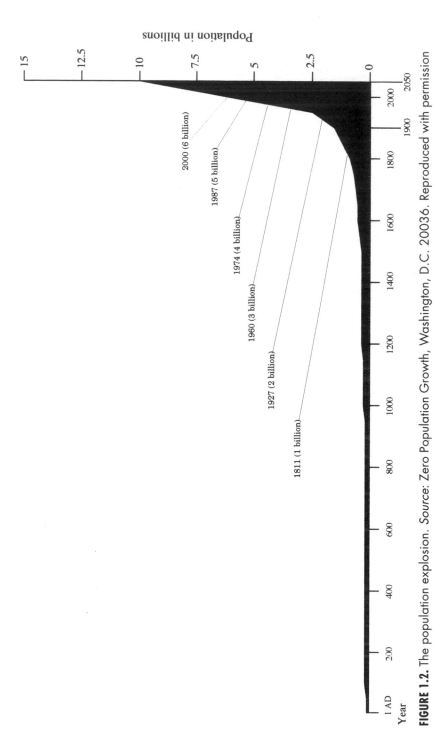

FIGURE 1.2. The population explosion. *Source:* Zero Population Growth, Washington, D.C. 20036. Reproduced with permission

creasingly expensive for communities to treat and dispose of this waste as specified by the law. Because of "not-in-my-backyard" (NIMBY) reactions of citizens, even siting a waste treatment or disposal facility is a major and expensive effort. Americans have 5% of the world's population and use about 25% of the earth's resources and about 25% of its energy. Through births and immigration, the population of the United States population is projected to grow by another 100 million in the coming 40 years and possibly double in the next 60. Beyond the United States, barring catastrophe, the world's population may double by the mid-twenty-first century. Figure 1.2 shows the steep population growth of the world in the recent past and projects even steeper growth in the coming century. These people will also understandably demand a better material life. Assuming that resources continue to be available to meet rising expectations, more pollution will result from the production of more goods. Given the greatest goodwill about limiting pollution and the best control methods, pollution will remain an issue.

Even if ever-increasing population and consumption were not problems, consider the following illustration of the difficulties that we would face if we tried to control every pollution source: In 1986, Congress passed a right-to-know law requiring industry to report its emissions of 320 chemicals by means of the Toxic Release Inventory (TRI). In the first TRI in 1988, American industry reported releasing nearly 2.6 billion pounds of the 320 chemicals into the air, a number that shocked many. At the same time there was a nationwide study of chemicals that might contribute to acid rain. This study examined all VOC emissions, representing many more chemicals than the 320 reported in the TRI. The study's conclusion? About 47 billion pounds of VOCs was released to the atmosphere per year, many times greater than the 2.6 billion pounds reported through the TRI. What are the sources of this huge quantity? Part is from industries not required to report TRI releases, such as mining and electric utilities and the hundreds of thousands of small businesses, dry cleaners, printers, painting operations, motor vehicle maintenance shops, even bakeries. Sewage treatment plants and landfills also release VOCs. Perhaps most striking, about one-half of the 47 billion pounds is emitted by the motor vehicles that are driven billions of miles each year by many millions of individuals – you and me. Despite the fact that new cars emit up to 98% fewer pollutants than in 1970, there are many more people driving more vehicles more miles. Many of these drivers do not maintain their vehicles to run as cleanly as designed. Individuals contribute to VOC emissions in other ways as well, through the use of products such as charcoal grill starters, paints, aerosol sprays, and pesticides. The list could be extended indefinitely. Emissions from any one individual are small, but multiplied by 260 million people and 100 million households, they are significant.

BOX 1.5

The individual and pollution. Consider the following statements: "More and more we are realizing that a large part of remaining environmental problems comes from the cumulative impacts of our individual actions. These impacts are far more subtle and, for many reasons, harder to manage than the single smokestack or wastewater discharge" (Martha Kirkpatrick, Maine Department of Environmental Protection, 1995). "More than half the nation's water pollution problems spring from everyday actions" (National Geographic Society and the Conservation Fund report, 1995). "We know that further progress on many fronts, notably air and water quality, means getting a handle on diffuse sources of pollution that result from millions of people making countless individual choices. The pollution challenge today is not General Motors, it is the general public" (William Reilly, former U.S. EPA administrator, 1995). More examples of individual contributions to pollution will become apparent in later chapters.

Can we regulate the emissions of many thousands of chemicals from hundreds of millions of businesses, communities, and individuals? Consider that the U.S. EPA cannot even convince most homeowners to test for radon despite the fact that it is a known human carcinogen. Likewise, although motor vehicles are a major source of air pollution, many people are unwilling to test emissions from their automobiles or only do so resentfully. EPA is continuing to regulate more and more chemicals and more and more sources, but even if authorized by law to do so, could federal, state, and local environmental agencies monitor every small business operation and every home in the nation? Other approaches are needed to deal with pollution. More education, especially more effective education, is one approach. In recent years, EPA has begun emphasizing another major approach, pollution prevention. Pollution prevention – avoiding the creation of pollution in the first place – will be discussed in Chapter 2 and thereafter emphasized throughout.

DISCUSSION QUESTIONS

1. In 1996, William Stavropoulos, the chief executive officer of Dow Chemical Corporation (the top chemical maker in the United States), announced that Dow would spend $1 billion on environmental health and safety (EH&S) in the next 10 years. By 2005, it seeks a 90% reduction in work place illness and injury, a 50% cut in

chemical emissions and a 50% decrease in waste and wastewater generation per pound of product manufactured. Furthermore, Dow expects a return on this investment. In the May 6 issue of *Chemical & Engineering News*, Mr. Stavropoulos said "We have known for a long time that pollution is environmentally and economically wasteful. Now we are going to prove it." He stated that two waves of environmentalism have occurred in the United States. The first in the 1960s and 1970s was driven by environmental organizations, and the second in the 1970s and 1980s by command-and-control legislation. He believes a third environmentalism wave will be led by industry because it can "link it to business reality." He foresees industry shifting from protesting against environmental regulations to embracing prevention activities, and from needing government regulation to assuming corporate responsibility. "Poor EH&S performance represents a waste of precious resources, raw materials, capital, people and the support of our communities."

(a) Does it make sense to you that industry might lead a "new wave of environmentalism"? Explain.

(b) Do you believe that Dow Chemical can realistically expect a monetary return on its billion-dollar investment in EH&S? How?

2. Americans spend $140 billion a year to control and clean up pollution. To put this expenditure in perspective, consider that United States companies spend about $160 billion a year on advertising. Many environmentalists believe that it is advertising that fuels a "culture of waste" and is responsible for the fact that Americans use a quarter of the world's energy and resources. We buy many products regardless of need – extra clothes, several radios and TV sets – new products even if the old have not worn out, and, often, more than one home. The list is long.

(a) How does advertising affect your behavior?

(b) Do you think that you buy more because of advertising than you would otherwise? Explain.

3. Environmentalists argue that consumerism is incompatible with environmental sustainability.

(a) What is your understanding of the word *sustainability*?

(b) Is it compatible with consumerism?

4. E. U. von Weizsaecker, of the Institute for Climate, Environment and Energy in Germany, commented in a 1996 letter to the periodical *Environment*, "The American (or, for that matter, European) lifestyle is not sustainable: Per capita resource use is too high by roughly a factor of ten."

(a) If he is correct, do you believe that we can find ways to use resources ten times more efficiently than we do now, enough to continue our high levels of consumption? Explain.

(b) If we reduced our consumption even twofold, what do you believe would be the effect on the American economy?

(c) Are there factors that could mitigate adverse effects? Explain.

5. (a) List the wastes and pollutants that you and others in your
household produce. Consider locations such as the following as
you prepare your list: kitchen, bathroom, laundry, basement,
yard, and garage.

(b) What wastes do you produce during trips away from home?
Consider transportation, lodging, restaurant meals, and other
activities.

(c) What personal wastes do you produce at your work site?

(d) Which of all the wastes that you have noted could you quite easily reduce, and which would be difficult or imposssible to reduce?

(e) Now consider the wastes and pollutants that you generate indirectly, those from your grocery store, gasoline station, auto
maintenance shop, dry cleaners, doctor's and dentist's offices,
and retail stores where you shop. Add those produced by manufacturers of your clothing, car, and other possessions. Do you as
an individual bear any responsibility for these? Explain.

FURTHER READING

Clarke, D. 1995. The Elusive Middle Ground in Environmental Policy. *Issues in Science &
Technology* XI(3), 63–70, Spring.

Clay, R. 1994. A Continent in Chaos: Africa's Environmental Issues. *Environmental Health
Perspectives* 102(12), 1018–23, Dec.

Clay, R. 1995. What More of Us Means. *Environmental Health Perspectives* 103(12), 1092–95,
Dec.

Crone, H. D. 1986. *Chemicals and Society*. Cambridge: Cambridge University Press.

Lepkowski, W. 1994. Ten Years Later, Bhopal. *Chemical & Engineering News* 72(51), 8–18,
Dec. 19.

Moore, J. 1990. Lists and Numbers: Effects on Public Policy? *Health & Environment Digest*
4(2), 1–2, Mar.

O'Riordan, T., Clark, W. C., Kates, R. W., and McGowan, A. 1995. The Legacy of Earth Day,
Reflections at a Turning Point. *Environment* 37(3), 7–15, 37–42, April.

Raloff, J. 1996. The Human Numbers Crunch. *Science News* 149(25), 396–7, June 22.

Reilly, W. K. 1995. Is There Cause for Environmental Optimism? *Environmental Science &
Technology*, 29(8), 366A-9, Aug.

U.S. EPA. 1994. *Terms of Environment, Glossary, Abbreviations, and Acronyms*. EPA 175-B-
94–015, April.

2

POLLUTION PREVENTION

The end-of-pipe approach to controlling pollution described in Chapter 1 has been called "first-generation" thinking. The intention of Chapter 2 is to introduce a second-generation concept, *pollution prevention* (P^2). Instead of controlling a waste, means are sought to produce less of the waste or, better yet, produce none at all. P^2 is the preferred step on the waste management hierarchy shown in Figure 2.1. When P^2 cannot prevent the waste or pollutant from being produced, the preferred means of handling the waste is recycling or reuse, the second step in the waste management hierarchy. When a material cannot be recycled or reused any longer, it is treated to reduce its volume or toxicity. Finally, at the bottom of the hierarchy, when no other option is available, the waste is disposed of. Each step of the waste management hierarchy is discussed in this chapter. P^2 is tremendously attractive as a concept and in practice but cannot solve all the pollution problems of a complex industrial society. Thus, some concepts and practices that go beyond P^2 will also be discussed.

POLLUTION PREVENTION

End-of-pipe control means capturing the pollutant after it is formed, but before it is released into the environment. Conversely, P^2 means decreasing the amount of waste or pollution produced in the first place: It is source reduction. With less waste or pollutant produced, less needs to be captured. Resource conservation, or increasing the efficiency with which we use raw materials like energy, water, and other resources, is also P^2. Using less energy is P^2 because fuel is conserved and, at the same time, pollutant emissions that would have resulted from the production and use of the energy are not produced.

Some examples of how industry uses P2 are given in Tables 2.1 and 2.2. For a business to practice P^2 successfully, its top management must become committed to its success. At the same time, employees must be motivated to take

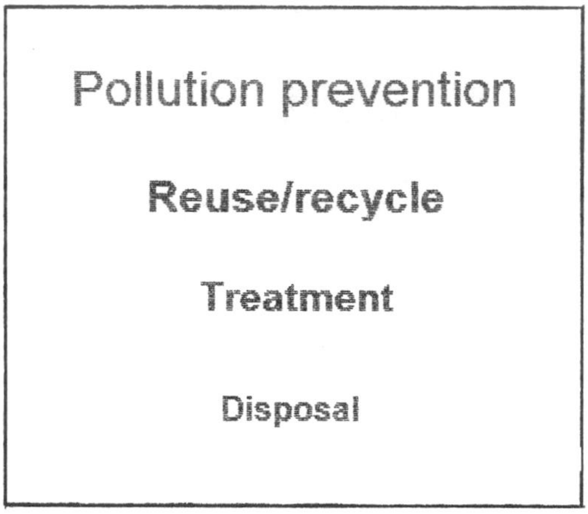

FIGURE 2.1. Waste management hierarchy.

P^2 seriously. As employees work with a process on a daily basis, they are often best informed as to where and how a process may be modified to decrease the waste generated or pollutants emitted. Some companies – 3M Corporation is a prominent example – offer bonuses to employees who develop workable P^2 ideas. The first step that a company often takes to begin its practice of P^2 is to pay attention to housekeeping practices. For example, if a hazardous chemical is spilled, it becomes a hazardous waste and the materials that must be used to clean up the spill also become hazardous waste. Finding means to minimize spills is P^2. Broader goals of P^2 include using less water and energy and fewer other resources to make products, reducing pollutant emissions during manufacture, and producing less waste, including hazardous waste. Reducing worker exposure to toxic chemicals during product manufacture, toxics use reduction, is another P^2 goal.

Some P^2 practices, such as improved housekeeping procedures, may be developed without great difficulty. Others, such as changing from an organic solvent to a water solvent, may change only one step in a production process and be implemented without great difficulty. To be most effective, P^2 involves design for the environment (DfE). In DfE, from the moment that a new product is conceived, it is designed with the environment in mind. DfE aims to develop a product that has a longer life, fewer environmentally harmful effects, and easier disassembly at the end of its life for reuse or recycling. EPA has a program that works with businesses on DfE efforts. Beyond the design of one product is the redesign of a whole production process, which can be complicated and costly. There is also the risk that a redesigned process may not be as effective as the earlier one. Some risk is definitely involved for the company using P^2.

The only U.S. federal legislation directly addressing P^2 is the 1990 Pollution Prevention Act, which uses the TRI report as a means to encourage P^2 prac-

TABLE 2.1 Examples of industrial use of P²

PREVIOUS ACTION AND RESULT	NEW ACTION AND RESULT
A company used an organic solvent, trichloroethane, to clean metal. Once dirty, the solvent became a hazardous waste.	The company found it could eliminate the trichloroethane cleaning step without ill effect on the product.
One company painted steel joists by dipping them into open vats of paint, releasing large amounts of fumes.	Vats were layered with ping pong balls, which greatly cut emissions without interfering with the dipping of joists.
A facility had storage tanks containing volatile organic chemicals. Large amounts of fumes were escaping.	Floating roofs were placed on the tanks. Because they float directly on the liquid, there is no room for vapor to form. VOC emissions to the air were greatly cut.
A soap manufacturer packaged its detergents in attractive boxes, often too large for their contents. The consumer threw away the packaging.	The manufacturer designed packaging that used less material, but that was still strong and attractive. The consumer throws away a smaller amount of packaging.
Manufacturers of beverage cans used protective thick aluminum.	Cans were redesigned to maintain strength, while using a third less aluminum. Less pollution was produced in making and transporting the cans.
An electric power plant used end of pipe controls to capture the sulfur dioxide that was formed as coal was burned.	The plant began buying coal that had been treated to remove much of its sulfur, resulting in lower sulfur dioxide emissions.

TABLE 2.2 P² in the motor vehicle industry

A MANUFACTURER DESIGNS A VEHICLE THAT	THE RESULT IS
Has reduced tailpipe emissions	Reduced air pollution
Has better fuel mileage	Reduced air pollution Reduced use of gasoline (conservation)
Is lighter weight, but still safe	Reduced air pollution Reduced use of gasoline
Uses fewer toxic chemicals in its manufacture	Reduced worker exposure to chemicals Reduced hazardous waste generation
Can be disassembled at the end of its life	Reduced solid waste generation Reuse of vehicle component parts

TABLE 2.3   Individual examples of P²

YOU
Purchase durable consumer goods.
Repair a television set or appliance rather than throwing it out.
Purchase a car with good fuel economy and maintain it to keep that economy.
Buy products with as little packaging as possible.
Avoid buying batteries for purposes that don't need batteries.
Use energy-efficient light bulbs and turn them off when not in use.
Turn down the thermostat at night and turn off appliances that are not being used.
Practice water conservation in the home and yard.
Drink more water or beverages prepared at home and fewer containerized beverages.
Eat less meat because less land, energy and water, and fewer pesticides and fertilizers, are used when grain is eaten directly than when it is used to feed livestock and poultry.

tices: It does this by requiring a business not only to report its chemical releases, but to describe what it is doing to reduce those releases. The EPA and state governments are also working with industry to stimulate voluntary P² efforts. In one effort, the 33/50 Program, EPA worked with industry to reduce emissions of 17 chemicals that are either very toxic or emitted in especially large amounts. The goal was reduced emissions of these 17 chemicals of 33% by the end of 1992 and 50% by the end of 1995. This goal was exceeded by the 1,150 participating companies. In the Green Lights Program, EPA works with business to reduce the amount of energy used in lighting. In the Energy Star Computers program, EPA works with desktop computer and laser printer manufacturers to develop products that go into an energy-saving power-down mode when not in use. P² is used not only by industries, but by municipalities, institutions, and individuals as well (see Table 2.3).

RECYCLING AND REUSE

Most people understand the importance of recycling and are familiar with recycling paper, glass, and aluminum and steel cans. The advantages of recycling aluminum are dramatic. Recycling saves 95% of the energy that would otherwise be needed to mine aluminum and make new containers. It also reduces the air and water pollution produced by mining aluminum by about

BOX 2.1

One small company's experience with pollution prevention. Over a 5-year period, a 170-employee tool and die manufacturing operation in Gorham, Maine, carried out a pollution prevention program, which accomplished the following: It cut the amount of hazardous waste it generated from 58,000 pounds to zero: Whereas in 1990, it spent $60,000 on hazardous waste disposal, by 1995 it spent $2,500, and by 1996 nothing at all – that is, it no longer generated hazardous waste. It also cut its TRI emissions from 18,000 pounds to zero. Whereas in 1990 it was one of the top 10 polluters in southern Maine, by 1994 it was receiving environmental awards for its pollution prevention successes. Another benefit was peace of mind for company employees who had previously been concerned about the chemicals they worked with and their company's poor reputation. It accomplished all this at a cost of $6,000. The strategy this firm, now owned by American Tool Companies, used was a step by step examination of the manufacturing steps that generated hazardous waste or polluting emissions. For each process of concern, it also asked its employees a question, Why do we do it that way? If the answer was, We've always done it that way, that was a signal to examine the process. Pollution prevention opportunities were analyzed with full employee involvement, never from the top administration down. Corrections were made step-by-step, pausing after each change to see how it was working. All possible sources of help were used including vendors, the state Department of Environmental Protection, and Maine's Environmentally Conscious Manufacturing Program.

95%, and aluminum resources are conserved. Although recycling savings for other metals are not as dramatic as for aluminum, energy and resource savings and reduced pollution are still significant. See Table 2.4 for other examples of industrial recycling and reuse and Table 2.5 for examples of individual recycling and reuse. Reusing a product is usually closer to P^2 than is recycling because fewer resources are used and less pollution is produced. For example, refilling glass bottles takes about one-third less energy than recycling the bottles. However, glass is heavy and the energy saved may be lost if the bottles have to be transported to a distant market.

DISCUSSION QUESTIONS

1. You need to explain to a group of grade school children the difference between pollution prevention and recycling.
   (a) How will you do it?

TABLE 2.4 Examples of industrial recycling and reuse

A factory uses a metal cleaning solvent. It reuses the solvent several times. When the solvent does became dirty, it is purified and recycled within the plant. It never becomes waste.

Instead of discarding its empty wire spools, a welding plant returns them to the supplier for reuse. The spools do not become waste.

A manufacturer uses recycled materials in the products it produces, and makes sure that the products it produces can be recycled.

A company finds uses for many of the wastes it produces.

An industry develops new uses for discarded consumer plastic. It also works to develop effective means to collect and identify used consumer plastics.

A paint manufacturer starts a plant to reprocess waste household paint.

An oil refinery starts a facility to re-refine waste motor vehicle oil.

TABLE 2.5 Individual examples of recycling and reuse

YOU
Make sure recyclable household items are recycled and not placed in the trash.
Find an individual or organization to use your leftover paint rather than disposing of it.
Take used oil to a service or recycling station.
Compost your grass clippings and leaves rather than placing them in the trash.
Reuse paper and plastic bags as long as they remain intact before discarding them or buy fabric bags, which can be reused indefinitely.

(b) Now think about reusing as compared to recycling. Using shopping bags as an example, explain how reuse is closer to pollution prevention than recycling.

2. You also need to explain to the children how to limit pollution when using batteries.
   (a) What is saved by using a rechargeable as compared to a throwaway battery?
   (b) What pollution is involved in recharging the battery?
   (c) Where might a rechargeable battery fit into the waste management hierarchy?

BOX 2.2

German recycling. Germany has a tough recycling law that includes a program to collect packaging materials. Indeed, so much packaging was collected in the early 1990s that Germany alone could not recycle it all. The excess was shipped to other European Union (EU) countries, interfering with their recycling programs. EU countries now have a rule that member states cannot set recycling targets far in excess of what they can handle within their own borders. Meanwhile, the German law is having the desired effect – the amount of packaging used is going down, 4% in 1993 alone. German manufacturers are now also packaging more products in glass and paper, which are readily recycled, rather than in harder-to-recycle plastic.

TREATMENT

Metals or glass can be recycled many times, but there is a limit to the number of times that paper or plastics can be recycled, and they eventually become waste. In addition, the recycling process is not waste-free, and a great deal of waste is sometimes produced during recycling. For example, when paper is recycled, paper coatings and fillers must be recovered. Because these do not have a use, they are disposed of in a landfill.

Once a material, recycled or not, becomes waste, the next step on the waste management hierarchy is treatment to reduce its volume or toxicity. A common means of treating municipal solid waste (MSW) or hazardous waste is combustion. Combustion of MSW is considered a treatment because it reduces waste volume, but some believe that combustion is not a treatment because the resulting ash contains higher concentrations of metals than the original waste. Combustion also produces at least some atmospheric emissions. Many states mandate the removal of lead acid batteries from MSW before it is combusted, to reduce ash toxicity and to reduce lead emissions to the air during burning. Partial recovery of other metals from MSW for recycling before it is burned reduces metallic air emissions during burning and reduces toxicity of the ash produced. However, many metallic materials remain in MSW.

Industrial combustion of hazardous waste is so heavily regulated that it can be cleaner than that of MSW. Consider dioxin emissions. Dioxins are a family of polychlorinated chemicals, several of which are especially toxic, in particular, 2,3,7,8-tetrachlorodibenzo-*p*-dioxin (TCDD, or dioxin). Emissions of even tiny amounts of dioxins are taken seriously. In 1994, EPA reported that combustion emits almost 8,500 grams of dioxin each year in the United States. Of this, only 35 grams is from hazardous waste incineration compared to 8,100 grams from medical waste and MSW combustion. However, EPA has now enacted much stricter regulations for MSW and medical waste combustors, and their future emissions of dioxin are also expected to be much smaller. Those combustors that now emit most dioxin and heavy metals to air will be required to reduce emissions by 99%.

There are many other means of treating hazardous waste in addition to incineration, and industrial hazardous waste cannot be legally disposed of unless it is first treated to reduce the hazard to specific standards. Examples of hazardous waste treatment are the neutralization of waste acid or waste alkali and the precipitation of metals from an industrial wastewater stream. Hazardous waste treatment will be further discussed in Chapter 8.

DISPOSAL

Disposal is at the bottom of the waste management hierarchy. Nonetheless, the way wastes are disposed of is very important. Materials that cannot be recycled or reused become waste, and recycled materials eventually become wastes. Sludge recovered from wastewater treatment can sometimes be recycled by composting, but often the only option is disposal. Solids recovered from air pollution control equipment also ordinarily become waste, although there is increasing effort to find uses for them. Even with increased recycling, most household trash becomes waste. Individual citizens can obtain information from a town office or library to aid responsible disposal of wastes that they produce. For example, waste paint that cannot be passed on to another user should be dried out (safely away from people or animals) before it is placed in the trash. Empty pesticide containers can be wrapped in paper before they are

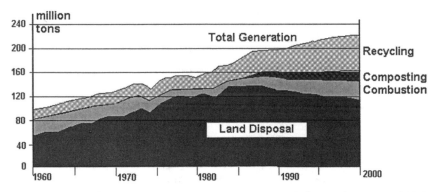

FIGURE 2.2. Trends in waste generation, recovery, and disposal. *Source: MSW Factbook*, Ver. 3.0, Office of Solid Waste, U.S. EPA, Washington, D.C., 1996.

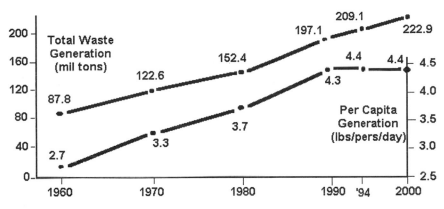

FIGURE 2.3. Waste generation rates, 1960–2000. *Source: MSW Factbook,* Ver. 3.0, Office of Solid Waste, U.S. EPA, Washington, D.C., 1996.

added to the trash. Used antifreeze can be taken to a recycling or disposal facility. Used oil is taken to a service station or other location where it can be recycled or properly disposed; it is never disposed of in a home drain or storm drain.

A snapshot of waste generation

Figure 2.2 shows the trends in MSW generation since 1960, and how recycling, combustion, or land disposal has been used to manage it. The United States recycled only a small percentage of its MSW in 1960. In the mid-1980s this figure began to increase, to reach 17% per year in 1990 and 22% per year in 1993. The 22% includes composting. In 1993, about 16% of MSW was combusted. Land disposal of MSW has slowly but steadily decreased since the mid-1980s. In 1993, about 62% of MSW was land-disposed as compared to 83% in 1985. Figure 2.3 could be interpreted as showing a somewhat more discouraging picture of waste generation. Although the waste generated per person has stabilized at about 4.4 pounds per person, the total waste generated continues to grow because of an increase in population.

DISCUSSION QUESTIONS

1. Instead of a flat fee to haul a household's waste away, regardless of the quantity, some communities charge householders per bag of garbage generated.
   (a) If your community began a per bag charge, what steps could you take to reduce your waste generation?
   (b) What steps would you actually be likely to take?

GOING BEYOND POLLUTION PREVENTION

Even with the best input of P^2, goodwill, money, and technical skills, some wastes will be produced. One question that could be asked is, Are there other approaches not immediately obvious from the waste management hierarchy that a company can take? One approach is for a business to look beyond its own operation and consider working with other companies. At the simplest level, this can mean listing its wastes (by-products) on a waste exchange, that is, marketing them to other businesses that can use them. However, such marketing is currently used to only a limited extent and does not usually involve systematic planning or cooperation with other businesses. Compare this human approach to the natural world. All living organisms depend upon other living organisms. Furthermore, living organisms use each other's wastes. Bacteria can degrade wastes produced by other living organisms and also degrade the bodies of those creatures when they die. In the process, the bacteria provide nutrition for themselves. Plants use an animal waste product, carbon dioxide, as a necessary nutrient and plants also use the organic material in soil – organic material from once living organisms – as a source of nutrients. Animals feed on plants to obtain calories and needed nutrients. What if industries could develop an analogous system of industrial symbiosis? An early example of industrial symbiosis is described in Box 2.3.

BOX 2.3

Industrial symbiosis. Kalundborg, Denmark, as described in a 1996 *Technology Review* article, provides an example of industrial symbiosis. Kalundborg has four industries, the Asnaes power station, which produces electricity by burning coal; Statoil Refinery, which makes petroleum products; Novo Nordisk, which manufactures pharmaceuticals and enzymes; and Gyproc, which produces wallboard. Over a period of 25 years, the following relationships have evolved.

Asnaes (by-products of a coal-burning electric power plant)

- Steam: Recall that only a portion of the energy generated by burning coal can be used to generate electric power. The rest ordinarily becomes thermal pollution. However, Asnaes uses its low-grade steam to provide heat to 5,000 houses and buildings in Kalundborg, to warm fish farms that it operates, and to provide steam to Statoil and Novo Nordisk. To do this, Asnaes is connected by a series of pipes to the steam recipients.

- Sulfur dioxide: Coal-burning power plants produce the pollutant sulfur dioxide from the sulfur in coal. Asnaes uses a scrubber to capture part of the sulfur dioxide formed and to convert it to calcium sulfate. It sells the calcium sulfate to Gyproc, the wallboard maker. Gyproc previously had bought its calcium sulfate from mines in Spain, but now Asnaes meets two-thirds of its needs.

- Coal-burning residues: Asnaes produces fly ash and clinker. Instead of sending these to a landfill, it sells them for use in building roads and in producing cement.

Statoil (by-products of a petroleum refinery)

- Burnable gases: Oil refineries often burn off the gases methane and ethane from petroleum. Not Statoil, which pumps its gas to Gyproc, which uses it to fire ovens to dry the wallboard. Gyproc needs only to have a butane gas backup system for the times Statoil shuts down for maintenance. Statoil also supplies gas to the Asnaes power plant, which can thus burn less coal.
- Sulfur: The process that Statoil uses to remove sulfur from the gas that it supplies to Gyproc and Asnaes produces a hot sulfur liquid that is shipped 50 miles away to Kemira, a sulfuric acid producer.

Novo Nordisk (by-products of a pharmaceutical firm)

- Sludge: Novo Nordisk uses microorganisms to ferment food-grade material to usable products. The residue remaining is a nutrient-rich sludge. The company uses steam from Asnaes to kill surviving microorganisms in this sludge. It then distributes the sludge by a pipeline network to a thousand nearby farms, which use it for fertilizer.

Another cooperative effort

- Water: To reduce the need for fresh water, scarce in Kalundborg, Statoil pipes its used water to Asnaes, which uses it to clean plant equipment and to provide feed water to its boiler, further reducing its thermal pollution. Overall, town industries have reduced water demand 25%.

DISCUSSION QUESTIONS

1. A manager of a large Maine paper mill commented that a business can save and make money once it gets "beyond compliance," that is, after it has fulfilled the requirements of the law to control pollution and then superseded those requirements.

    (a) How does this mill save money by using energy more effi-
ciently?

    (b) By using water more efficiently?

    (c) By recovering paper fiber, which previously went to the sewer?

    (d) By reducing the amount of solid waste going to a landfill?

    (e) By finding markets for materials previously disposed of in a
landfill?

    (f) By largely eliminating hazardous waste generation?

    (g) It is unusual for a manufacturing facility to recover the green-
house gas carbon dioxide. However, this mill is recovering 10%
of its carbon dioxide and converting it into calcium carbonate,
a paper-making chemical. Is this pollution prevention or pollution control, or can it be completely explained by either concept?

    (h) What are possible pollution prevention activities that this mill
or other facilities might find more difficult to implement and
might not save money?

2. Review the Kalundborg example. Consider the industries in your
own or another community.

    (a) What possibilities do you believe may exist for industrial sym-
biosis in these industries?

    (b) What factors – physical location, financial considerations, liabil-
ity, and so on – could favor the development of symbiosis?

    (c) What factors could limit it?

    (d) Could the community participate in the symbiosis? How?

FURTHER READING

U.S. EPA. 1994. Voluntary Programs. *Pollution Prevention News*, June-July EPA 742-N-94–003:
2–5.

Browner, C. M. 1993. Pollution Prevention Takes Center Stage. *EPA Journal* 19(3), 6–8, July-
Sept.

Gertler, N., and Ehrenfeld, J. R. 1996. A Down-to-Earth Approach to Clean Production.
Technology Review 50–4, Feb./Mar.

Hirschhorn, J. S., and Oldenburg, K. U. 1991. *Prosperity Without Pollution*. New York: Van
Nostrand Reinhold.

3

AN INTRODUCTION TO TOXICOLOGY

BACKGROUND

"All substances are poisons. There is none which is not a poison. The right dose differentiates a poison and a remedy." This statement, made by the sixteenth-century physician Paracelsus, is still quoted today, 450 years after his death. It states a central premise of toxicology: If the dose is high enough, any substance can be harmful to living creatures. The opposite also holds true: If the dose is low enough, even extremely toxic substances may not have an adverse effect. See Table 3.1 for definitions of terms used for toxic substances. Although the term *toxic chemical* is sometimes often used almost as one word, the toxicity of a chemical is seldom so simple. Effects of a chemical depend on many factors, including dose, health, age, and sex of the person or animal exposed and the specific conditions of exposure. Table 3.2 provides some examples that demonstrate this relationship.

Acute and chronic toxicity

Acute toxicity and chronic toxicity are often distinguished. *Acute toxicity* is an adverse effect seen soon after a one-time exposure to a chemical. Examples of acute effects are vomiting, diarrhea, irregular heartbeat, lack of coordination, and unconsciousness. Some of these symptoms might be seen in a child who ingested a parent's prescription drug, a farm worker who sprayed a pesticide without proper protection, or a teenager who sniffed glue or gasoline vapors. In contrast, *chronic toxicity* results from long-term exposure to lower doses of a chemical. A major example is cancer, which usually does not develop until long after an initial exposure to a substance. Leukemia, a cancer of white blood cells, may result from long-term exposure to benzene, or lung cancer may result from long-term cigarette smoking. Toxic effects on the nervous system may result from chronic exposure to mercury, or liver cirrhosis may result from chronic alcohol ingestion. Some substances that may not cause acute

TABLE 3.1 Definitions

Toxicant	A substance that can cause adverse effects in a plant, animal, or human.
Toxin	A toxicant produced by living organisms, microorganisms, plants, insects, spiders, or snakes. In common usage, the word *toxin* is used as a synonym for *toxicant*.
Poison	Legally, a poison is a substance that is fatal at a dose of less than 50 mg/kg body weight. Like the word *toxin*, *poison* is often used loosely to represent any toxicant.
Hazardous	A hazardous substance may be toxic, corrosive, reactive, flammable, radioactive, or infectious, or some combination of these.
Xenobiotic	A chemical foreign to (not synthesized in) the body of animal exposed to it.

effects may cause chronic effects. If you break a mercury thermometer, your onetime exposure to elemental mercury vapor is unlikely to cause adverse effects. However, if you suffer chronic exposure to mercury vapor in a work place, this can lead to serious effects on the nervous system. It is also true that some substances that cause acute effects may not show chronic toxicity. The foul-smelling gas hydrogen sulfide, is very acutely toxic. However, long-term exposure to low doses of hydrogen sulfide found naturally in "sulfur waters" is not known to have adverse chronic effects. Indeed, in past years, exposure to these odoriferous waters was considered healthful.

Dose

Anything can be toxic if the dose is high enough. Even drinking very large quantities of water has killed people by disrupting the osmotic balance in the body's cells. Figure 3.1 shows a dose-response curve, the increasing effect that is seen with increasing dose. If the dose is low enough, most chemicals show no adverse effect. However, as the dose is increased, an enzyme may suffer increasing loss of its activity or nerves may become unable to conduct impulses. Table 3.3 lists several examples of how toxicants exert their adverse effects. How toxic a chemical is as compared to other chemicals is measured by noting what dose is lethal to half the animals exposed to it, the LD_{50} (see Table 3.4). Determining the LD_{50} is crude and many consider it cruel as well. Furthermore, it does not provide good information as to how a chemical exerts toxic effects. However, people continue to want information on the relative toxicity of a chemical and so these studies continue.

TABLE 3.2 Toxic or beneficial?

Botulinum toxin	A toxin produced by bacteria in improperly processed food. As the most acutely toxic chemical known, it has caused many human deaths. In very tiny doses it is a medical treatment for muscular spasms and twitching that have responded to no other treatment.
Warfarin	A rat poison that prevents blood clotting. It is used to prevent strokes and heart attacks.
Atropine	A 'supertoxic' chemical produced by the deadly nightshade plant. It is used as an antidote for nerve gas and organophosphate pesticide poisoning.
Thalidomide	A drug that caused tragic birth defects in the 1960s. It is used as a leprosy treatment, may be used in treating acquired immune deficiency syndrome (AIDS), and is being studied as an immune system suppressant in people receiving organ transplants.
Curare	A poison used on arrow tips by Amazon natives. It is used to promote muscle relaxation during surgery.
Nitroglycerine	A chemical used to manufacture dynamite that is employed to treat angina (spasms of heart arteries).
Sodium chloride (salt)	An essential nutrient. It has killed small children who have ingested an excess. Ingestion of too much salt is implicated in high blood pressure and retention of body fluids. Long-term consumption of highly salted foods is associated with stomach cancer.
Vitamin A	An essential nutrient. Worldwide, millions of children's deaths are attributed to insufficient vitamin A intake. High doses taken by pregnant women are associated with serious human birth defects.
Nickel and chromium	Both are essential nutrients. High doses are toxic and can cause cancer.
Nitric oxide	A neurotransmitter produced within the body. It is an ambient air pollutant.
Chlorinated dioxins	Among the most toxic chemicals known. In test animals, they powerfully affect development and the immune system and cause cancer. They are being studied as a treatment for breast cancer.

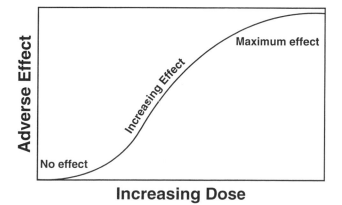

FIGURE 3.1. Adverse response as a result of increasing dose.

BOX 3.1

The paradox of aspirin. Common aspirin has a fascinating range of effects. It relieves pain, reduces fever and the inflammation of arthritis, treats and prevents heart attacks and strokes, and may prevent colon cancer. This very same aspirin is also a stomach irritant and may have adverse effects in children with fever or in individuals with blood-clotting problems. It can cause bleeding in pregnant women and is highly toxic to those allergic to it. Aspirin also causes birth defects in rats, and, if it were a new product, it probably would not be sold as an over-the-counter drug. This is true because modern drugs are extensively tested before marketing. If aspirin were a new substance and testing had demonstrated that it caused birth defects in animals, it would be suspected of doing the same in humans. If available, it would be marketed as a prescription drug. However, aspirin has been on the market since the nineteenth century and has not been associated with human birth defects.

Figure 3.2 shows a dose-response curve for an essential nutrient, a vitamin, mineral, or amino acid. For a nutrient, there are adverse effects when the dose is too low. As the nutrient dose is increased, the animal responds positively and continues to do so through an optimal dose range. At yet greater doses, adverse effects are again seen, and, if the dose is high enough, serious illness or death may occur.

TABLE 3.3   Examples of ways that a toxicant can exert adverse effects

Carbon monoxide	The purpose of the blood protein hemoglobin is to pick up oxygen in the lungs and transport it to the body's tissues, where the oxygen is released. Carbon monoxide binds to hemoglobin much more strongly than oxygen. In so doing, it blocks the binding of oxygen, and lowers the amount of oxygen available to the body. If enough hemoglobin is blocked, death results. Lesser amounts of carbon monoxide can cause headache, nausea and other flu-like symptoms and are also implicated in the development of heart disease.
Botulinum toxin	This powerful toxin binds to nerve endings at points where they join muscles, and blocks release of the neurotransmitter, acetylcholine. This results in muscular paralysis. The immediate cause of death is usually paralysis of respiratory muscles.
Nerve gases and organophosphate pesticides	These agents prevent acetylcholine from being degraded. Thus, acetylcholine accumulates and results in uncontrolled firing of the nerves.

TABLE 3.4 Comparative chemical toxicity

DEGREE OF TOXICITY	LD$_{50}$[a]	EXAMPLES
Slightly toxic	500-5,000	Aspirin Vanillin Salt
Moderately toxic	50-500	Phenobarbital Caffeine Nicotine Warfarin
Highly toxic	1-50	Sodium cyanide Vitamin D Parathion
Supertoxic	< .01-1	Atropine Nerve poisons TCDD (dioxin)
Biotoxins	<< 0.01	Botulinum toxin Ricin (in castor oil beans)

[a] The LD$_{50}$, expressed as milligrams per kilogram (mg/kg) body weight, is the dose that kills 50% of the animals exposed to it.
Source: Adapted from H. D. Crone, *Chemicals and Society*, Cambridge: Cambridge University Press, 1986, p. 35.

Dose per time

The period of time over which the dose is received is as important as the dose itself. Taking one aspirin a day for 100 days may have a beneficial effect, but taking 100 aspirins at one time could be lethal. Ingesting the alcohol in 1 ounce of hard liquor every day for 25 days would not adversely affect most people, but drinking a fifth (25.6 ounces) at one sitting could be lethal. Or, although caffeine is moderately toxic, a person can safely drink coffee one cup at a time. However, the amount of caffeine in 100 cups of strong coffee could be lethal. Anyone who has chewed a piece of raw rhubarb is familiar with the effect of oxalic acid, a chemical also found in spinach, on the mouth. Ten to 20 pounds of spinach or rhubarb eaten at one sitting could provide a lethal dose of oxalic acid. The chemical solanine found in potatoes is toxic, but it would take about 100 pounds of potatoes eaten at one sitting to provide a lethal dose. These last three examples may be amusing, but they do demonstrate the point that toxic substances are found in anything that we eat or drink and the dose makes the poison.

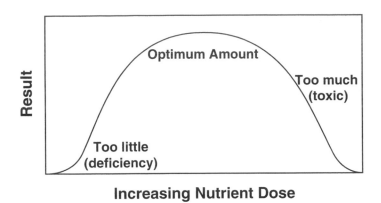

FIGURE 3.2. Dose–response curve for an essential nutrient.

ABSORPTION, DISTRIBUTION, METABOLISM, AND EXCRETION (ADME)

The acronym ADME may help you to remember what happens to a chemical as it enters and moves through the body. After a chemical gains entry to the body through the lungs, gastrointestinal tract, or skin, it is absorbed into the bloodstream and then distributed throughout the body, entering various organs. It is then metabolized (biotransformed) in the body's organs. Finally it is excreted from the body.

Absorption

For a chemical to have an effect, it must first contact the body: There must be exposure. A chemical ordinarily has three possible routes of entry into the body: inhalation into the lungs, ingestion into the gastrointestinal tract, or absorption across the skin. A chemical can also gain entry to the body by direct injection into the bloodstream or under the skin, but these pathways are not ordinary means of entry. Some chemicals are toxic only by one route of entry. Formaldehyde is a carcinogen only if inhaled. Radon is also primarily a carcinogen by inhalation. Arsenic is toxic by all three routes: skin absorption, ingestion, and inhalation.

Ingestion. Anything taken into the body by drinking or eating is ingested. Once in the gastrointestinal tract, a substance can be absorbed into the bloodstream from any point within the tract. However, most absorption is from the small intestine, a location in contact with a portal blood system, which carries the substance directly to the liver. The liver is the organ that receives the highest dose of ingested toxicants.

Inhalation. Apart from deliberate injection of a chemical into the bloodstream, inhalation is the fastest means by which it can enter the body and exert an effect. An example is the inhalation of a gaseous anesthetic such as ethyl ether. Anesthesia results very rapidly after the inhaled gas passes from the lung's alveoli into the bloodstream. Nongaseous substances can also be inhaled. Examples of these are fumes from hot metal and the particulates in smoke and in the spray from an aerosol can.

Skin Absorption. The skin stops or slows the absorption of most chemicals. However, if a chemical can penetrate the skin's epidermis and reach the dermis, it may be absorbed into the blood. A few chemicals like the pesticide parathion are absorbed as rapidly by skin absorption as by ingestion or inhalation. One unusual chemical, dimethyl sulfoxide (DMSO), is sometimes used as an agent to enhance absorption of chemicals that otherwise are not significantly absorbed across the skin. The thin skin found on the abdomen or scrotum is more permeable to a chemical than the thicker skin on the soles of the hands or feet. A larger area of skin exposed to a chemical absorbs more than a smaller area. Likewise, the longer a chemical is in contact time with the skin, the more that can be absorbed. Some chemicals cause local adverse effects on the skin, eyes, lungs, or gastrointestinal tract; that is, adverse effects are seen without absorption into the blood. However, the words *toxicant, toxin,* and *poison* most often refer to substances that have an effect distant from the point of intake, that is, systemic effects.

BOX 3.2

Plutonium exposure. There is concern, especially in countries of the former Soviet Union, that the radioactive element plutonium recovered from dismantled nuclear bombs will be stolen by terrorists and used to make bombs. The potential adverse effects of plutonium vary, depending upon the route of exposure. Most of the ionizing radiation that it emits is in the form of alpha particles. If plutonium is ingested or inhaled, alpha particles can have very damaging effects within the body. However, alpha particles do not penetrate the skin, so plutonium can be handled by someone wearing only a lead apron and it can be carried with little fear of personal injury. It has been strongly suggested that plutonium from dismantled bombs be melted into kiloton-size glass logs to make it much more difficult to steal.

Bioavailability. Whatever its route of entry to the body, a chemical must be bioavailable in order to be absorbed into the bloodstream. Pollutants such as dioxins, PCBs, and polycyclic aromatic hydrocarbons (PAHs) bind tightly to soil and other particles. For example, tightly bound chemicals are often inaccessible to microorganisms that might otherwise be able to degrade them. When such firmly bound pollutants are ingested, only a portion, the bioavailable portion, may be absorbed into the bloodstream from the intestine or the lung. Many factors affect the bioavailability and thus the toxicity of a pollutant.

Physical and chemical factors also affect how much of a chemical will be absorbed into the body and how fast it will be absorbed. For example, not all chemical forms of a metal are equally well absorbed across the intestinal wall. If a person accidentally ingests elemental mercury, little or none of this insoluble metal can be absorbed and it passes through the gastrointestinal tract. However, some compounds of mercury can be absorbed across the intestinal wall into the blood. The fat-soluble methylmercury is especially well absorbed. Sometimes, differences among chemicals can be used to advantage. For example, barium compounds are toxic; nonetheless, barium sulfate is safely used in X-ray diagnosis of the colon because it is insoluble both in water and in fat. When ingested, it passes through the body and is excreted. However, barium chloride could not be used this way because it is soluble enough in water that a portion would be absorbed.

Distribution

After a chemical is absorbed into the blood, it is distributed throughout the body and taken up, to varying extents, by different organs. A specific chemical often has a greater effect on one organ, the target organ, than on others. To exert a toxic effect, a chemical must reach a sensitive organ at a high enough dose to exert an adverse effect, and it is this dose that is important. Bone marrow is the target organ of the petrochemical benzene. The nervous system is the major target organ of the heavy metal lead in a fetus or young child. However, the existence of a target organ does not mean that other organs are not affected. As dose increases or time of exposure increases, additional organs may be affected.

Storage. A portion of the chemical distributed around the body may be stored. Even the blood can store chemicals if they are bound to blood proteins. As long as a chemical remains bound in its storage spot, it does not ordinarily exert adverse effects. An exception to this protective effect of storage occurs with radionuclides, radioactive chemicals. These can decay and cause potential harm, whether they are stored or not. For example, the radioactive chemical strontium-90 was found in the fallout of nuclear bombs tested in the atmosphere in the 1950s. Absorbed into the body, strontium-90 is stored in bones, where it increases the risk of bone cancer.

A word commonly used to refer to storage of a pollutant at levels higher than those found in the environment is *bioaccumulation*. PAHs, PCBs, dioxins, and organometallic forms of metals bioaccumulate in fat. Fluoride and lead bioaccumulate in bones. Chemicals bound to proteins, cadmium, for example, can bioaccumulate in the liver, kidney, and other soft tissues. The term *biomagnification* is similar to *bioaccumulation* but refers to the fact that certain contaminants bioaccumulate in increasing amounts moving up the food chain. Methylmercury, found in the sediments of a water body, is an important example of a pollutant that displays this behavior. Tiny invertebrate animals are the first organisms in the food chain to absorb and concentrate methylmercury from sediments. Small fish that eat these animals concentrate the methylmercury to a greater extent. Larger fish eating the smaller fish concentrate it even more. Finally a bird or mammal – examples are eagles, gulls, seals, minks, or human beings – eats the larger fish and the methylmercury concentration increases further. The polychlorinated chemicals, PCBs, DDT, and dioxins are other well-known examples of chemicals that undergo biological magnification.

When exposure to a chemical is reduced or eliminated, the amount of chemical stored in the body decreases. It may be released from its storage site slowly under ordinary conditions; however, some conditions result in rapid release. Lead is ordinarily released very slowly from its storage place in bones. However, during pregnancy, a mother's bones release more calcium than usual to meet the needs of her fetus. At the same time, larger than usual amounts of lead are released and it is this higher dose of lead to which the fetus is exposed. Another example is DDT, which is usually only slowly released from fat. In one experiment, laboratory rats were fed high amounts of DDT for 3 months. Although the DDT accumulated to high levels in their fat, the rats showed no ill effects. However, when the food intake of the rats was cut in half, the animals were forced to use stored fat for energy. The stored DDT was released rapidly at the same time and the rats showed visible symptoms of poisoning. Wild animals also go through periods when they use up large amounts of body fat, for example, during periods of famine or when nursing mothers use body fat for the extra energy needed to produce milk. At these times, pollutants stored in fat may be rapidly released and potentially present problems.

Metabolism

The word *metabolism*, sometimes called *biotransformation*, describes how living organisms convert absorbed chemicals into other chemicals. All organs carry

BOX 3.3

An environmentally persistent and ubiquitous family of chemicals. PCBs are a family of 209 chemicals, each varying in the number and location of chlorine atoms that it contains. Each also varies in toxicity. About 40 to 70 of the 209 were found in industrial mixtures that were manufactured over 40 years and used for many purposes. Major amounts of PCBs were used to insulate and cool electrical equipment. Because of their environmental persistence and their demonstrated toxicity to wildlife, PCB manufacture was banned in the mid-1970s. By this time, several hundred million pounds (of about 1.4 billion pounds manufactured) had been released to the environment. Large amounts were accidentally spilled on land in locations where they were used in electrical equipment or were deliberately disposed of at waste sites. As a result, PCBs are the primary pollutant of concern at 20% of the nation's high-priority hazardous waste (Superfund) sites. Many waterways were also contaminated. Parts of New York's Hudson River suffered especially heavy PCB contamination, as did many locations along the Great Lakes and marine coasts. Because of their low water solubility, PCBs concentrate in sediments. From this sediment sink, they are taken up and concentrated in living organisms. Fish consumption is the major route of human exposure to PCBs. At levels found in the environment, PCBs do not ordinarily manifest acute toxicity in humans so concern centers around potential chronic effects on reproduction or the immune system. Animals degrade polychlorinated chemicals very slowly, so they remain in body fat for many years. Other organisms also have difficulty in degrading PCBs. This includes the microorganisms that ordinarily degrade the organic chemicals found in dead plant and animal matter, which also degrade organic pollutants. But most microbes very slowly degrade PCBs and other polychlorinated chemicals. They have an especially difficult time degrading heavily chlorinated PCBs. It is these that are most environmentally persistent and concentrate most heavily in fat.

Environmental levels of PCBs are now much lower than they were in the 1970s, but they nonetheless remain ubiquitous environmental contaminants. In particular, many hot spots suffered heavy PCB contamination, and they remain a concern in the 1990s. PCB concentration in Great Lakes fish is as much as 90% lower than in the 1970s. Still, many fish exceed the PCB tolerance level set by the U.S. Food and Drug Administration (FDA), that is, the fish are considered unsafe to eat. As often happens with environmental pollutants, wild animals have more direct exposure than humans. PCBs have been found in the fat of Beluga whales in Canada's St. Lawrence estuary at levels as high as 600 ppm, a tremendously high level of contamination. The less than 5 ppm PCBs found in Arctic whales is still high. Remember that a parts per million (ppm) level is a thousandfold greater than a parts per billion (ppb) level and a millionfold greater than parts per trillion (ppt). Levels of parts per billion or parts per trillion are more typical of polychlorinated contaminants found in animals or humans.

A North Atlantic mystery. Since 1988, a number of serious epidemics have killed many seals and dolphins. In one epidemic, 20,000 seals, about half of the seal population of the North Atlantic and Baltic Sea, died as the result of a viral infection. These marine mammals have a very large amount of body fat (blubber), and some suspected they had become abnormally susceptible to infection because accumulated pollutants in their bodies had weakened their immune system. A Dutch scientific team carried out a study to test this suspicion by capturing a group of seal pups. Eleven control pups were fed fish from the relatively unpolluted North Atlantic Sea; the other pups received fish from the Baltic Sea that contained about 10 times more PCBs. Pups fed the contaminated Baltic fish showed immune system changes that could interfere with their ability to fight infection, changes not seen in the control pups. Natural killer cells (NKCs) are a population of immune cells that can destroy cells infected with virus. Samples of NKCs taken from pups fed Baltic fish indicated that they were 25% to 50% less active – and presumably also less active in combating infecting cells – than NKCs of control seals. The activity of a second population of immune cells, T cells, was also suppressed.

These results are consistent with the hypothesis that PCB contamination weakened the immune system of seals killed in the 1988 epidemic, but they do not confirm it. A more telling test would be to expose the pups to the virus that caused the 1988 epidemic to determine whether they were indeed more susceptible to infection. However, even where such tests are legal, most researchers would not want to perform them. Even without such an experiment, the results of this study raise serious concerns about the high PCB levels found in the Baltic Sea. The Baltic borders on a number of the countries of the former Soviet Union.

out biotransformation, but the liver is especially active. Biotransformation usually converts a xenobiotic into a less toxic chemical. Unfortunately, in some cases, the result is a more toxic chemical. For example, the liver converts the pollutant benzene into benzene oxide, a reactive chemical that can damage bone marrow. The biotransformation system is an ancient one that evolved to deal with xenobiotics. From this perspective, modern pollutants –

with exceptions like PCBs – are no worse than the toxins that animals have been dealing with for millions of years.

Excretion

Water-soluble chemicals are largely excreted from the body in urine. Some chemicals such as salt and sodium cyanide are water-soluble; others become soluble as the result of biotransformation, allowing them also to be excreted in the urine. Volatile chemicals like ethyl alcohol and acetone are partially excreted in the exhaled breath. You are probably familiar with the smell on the breath of a person drinking alcohol. Gases like carbon monoxide are also expired with the breath. Chemicals that the liver cannot transform into water-soluble forms are excreted with the bile into the intestine and are then excreted with the feces. However, a portion of these chemicals may be reabsorbed across the intestine into the blood and carried once again to the liver. Sometimes, such chemicals are cycled a number of times before being completely excreted. Fat-soluble chemicals can be excreted in the milk of a nursing mother, sometimes significantly increasing the exposure of nursing infants to chemicals like PCBs and dioxins. A small amount of chemicals is also excreted in sweat.

OTHER FACTORS AFFECTING TOXICITY

How toxic a chemical will be is affected by its inherent toxicity, its dose and dose per time, the quantity of it absorbed into the body, and whether it is converted to a more or less toxic form. Some other factors that affect toxicity follow.

- Species. A chemical that is toxic in one species at a given dose may have little effect in another. In the past, silicosis often developed in human miners exposed to silica dust, but in the mules exposed along with them it did not. Or, consider the extremely toxic chemical 2,3,7,8-TCDD. Its toxicity varies greatly, depending upon the animal species (see Table 3.5). Individuals within a species also vary greatly in their sensitivity to a chemical; some may be unusually sensitive to a certain substance and others resistant. Some humans, for example, are hypersensitive to aspirin or to the sulfites used to preserve wines and some foods. Health-based standards are usually set to protect the most sensitive individuals.
- Sex. A chemical may affect the ovaries of a female but not the testes of a male, or vice versa. The sexual hormones androgens and estrogens affect many aspects of anatomy, physiology, and metabolism, so it is not surprising

TABLE 3.5 Variation in dioxin toxicity
with species

SPECIES	LD_{50} (μg/kg body weight)
Guinea pig (male)	0.6
Rat (male)	22
Rat (female)	45
Hamster	1,160-3,000
Rabbit	115
Dog	30-300
Monkey	70
Humans (estimated)	>100

Source: Adapted from F. H. Tschirley, "Dioxin,"
Scientific American, 254(2), Feb. 1986, pp. 29–35.

that the sexes can have different reactions to chemicals. Women, for example, have in their bodies less alcohol dehydrogenase, which is the first enzyme involved in alcohol metabolism, than do men. The result is that a woman can become intoxicated more rapidly than a man of the same weight. Despite differences between the sexes, new drugs have ordinarily been tested only in men. Because women have protested their exclusion, this situation has recently improved.

- Age. Babies and small children have less well-developed immune systems than do adults, and they also eat more food per pound of body weight. So, if a food they eat is contaminated, children ingest proportionately larger amounts of the toxicant. Nonetheless, children are less sensitive to some toxicants than adults. At the other end of the age scale, elderly persons may have immune systems that no longer function well. Thus, the elderly may react more strongly to a drug or pollutant than younger adults.

- Nutrition. A poorly nourished animal or person is more susceptible to toxicants than is a well-nourished one. For example, children who ingest too little of the metal nutrients calcium and iron in their diet absorb more of the heavy metal lead, which may contaminate their food, than do children with optimal dietary calcium and iron intake. A poor diet can also have an adverse effect in adults. The dietary deficiencies of alcoholics make them more susceptible to toxic substances and infectious microorganisms. Those consuming a high-fat diet have a greater risk of colon and skin cancer. In fat rats or even rats of "normal weight" more cancers develop than in rats fed a well-balanced low-calorie diet. Heavier people have a greater risk of cardiovascular diseases and cancers, including postmenopausal breast cancer and cancer of the endometrium (the inner lining of the uterus).

DISCUSSION QUESTION

The metal cadmium is not a nutrient and can be toxic at low doses. Some years ago, a poisoning incident occurred in Japan among poor elderly women. The women affected had each had several children,

---

BOX 3.4

"Lead is not lead is not lead." Smuggler Mountain, Colorado, was an active mining site until early in the twentieth century. Mining operations left behind a large land area contaminated with lead and cadmium. In 1982, soil tests showed such high lead levels that the U.S. EPA put Smuggler Mountain on its list of Superfund sites – those hazardous waste sites that pose a special risk. EPA proposed a $12 million cleanup that involved excavating and removing contaminated soil. Small children are at special risk from contaminated soil as they are most likely to ingest it during play. When children have blood lead levels of 10 micrograms per 100 milliliters (µg/100 ml) or greater, the U.S. Centers for Disease Control and Prevention (CDC) recommend that exposure be reduced. But, despite high environmental lead levels, Smuggler Mountain children had blood lead levels that averaged 2.6 µg/dl.

Typically, communities near Superfund sites are anxious about the hazards they pose and want them cleaned up completely. However, because EPA was planning excavations that could destroy their community and because the children had low blood lead levels, citizens protested the cleanup plan at Smuggler Mountain. A technical advisory committee (TAC), an independent panel of experts, was formed to evaluate the situation and to make recommendations. The TAC noted a number of factors that eased concerns about the high lead levels in the soil. Only a few hot spots contained high lead levels. Residents had diets high in iron, which mitigated exposure to lead; the areas that were most heavily contaminated with lead and cadmium were covered with grass in summer and snow in winter, which reduced exposure to them. To the community's relief, the TAC recommended a much more limited remediation plan at a cost of only $400,000. One TAC member made the following observation: "Lead is not lead is not lead: trying to devise a single clean-up value for soil lead is not practical to use at all sites; each site must be analyzed individually."

and their diet consisted primarily of rice grown in paddies contaminated with cadmium; *itai itai* (pain-pain), a disease characterized by kidney damage and brittle, painful bones, developed. Other people who ate the rice were not adversely affected. What factors could have caused only these particular women to be adversely affected?

ORGAN SYSTEMS AFFECTED BY TOXICANTS

It is impossible to describe here all the organ systems that can be affected by toxicants, and the effects seen. The information that is presented serves only as an introduction to this topic.

Skin

Some chemicals have local effects upon contact with the skin. For example, a strong acid is a corrosive that can directly and seriously damage skin. An irritant chemical has a less severe effect, but causes skin reddening, swelling, or itching, which subsides after exposure ceases. Detergents and other cleaning products are skin irritants to some people, as is the metal nickel, often used in jewelry. Exposure to certain plants also irritates skin. If a person is exposed repeatedly to an irritating chemical, the skin reacts each time, but the irritation subsides when exposure ceases. A nonchemical irritant is sunlight, which can cause reddening, pain, and sensitivity. More serious problems can arise when an irritant is also an allergen. When a sensitive person is exposed for the first time to an allergenic chemical, irritation ceases when exposure ceases, as described. However, repeated exposures in sensitive people lead to reactions that grow in severity with each exposure and reactions that occur with much lower doses – an allergy has developed. Formaldehyde is a well-known irritant, which is also an allergen. Local reactions are very important when considering skin. However, certain chemicals affect skin only after they have been absorbed into the body, that is, they act systemically. The antibiotic neomycin is an example of an ingested chemical that results in skin irritation. Yet other chemicals – arsenic is an example – can adversely affect the skin through both local and systemic reactions.

Lungs

As with skin, some chemicals have local effects on the lungs. The reactive gases ozone, formaldehyde, ammonia, chlorine, and chlorine dioxide are among the chemicals that can directly damage the lungs. The reactive gas sulfur dioxide is converted into acid after contacting the moisture in the lungs and the acid damages the lungs. Reactive gases also affect the mucous membranes in the eyes and nose. Substances that are not chemically reactive can also damage the lungs; examples are silica, asbestos, coal or cotton dust, and even talcum powder that has been applied too liberally. The adverse effects can be either acute or chronic; for example, inhalation of various dusts can have an immediate irritant effect, but chronic exposure can damage lung function, as when cotton dust gives rise to "brown lung" disease over a period of years. Other substances are lung carcinogens, including asbestos, radon, and chemicals in tobacco smoke. Lungs can also be damaged by organic solvents, and accidental aspiration of gasoline can cause severe lung damage or death. In other instances, the lung serves as an entry point for chemicals that have systemic effects. Chemicals, especially volatile toxicants, enter the bloodstream through the alveoli, tiny air sacs deep within the lungs at the end of the lung bronchiole tubes. The normal function of the alveoli is to provide a surface for an exchange between oxygen and carbon dioxide, but other chemicals can follow the same route.

Liver

The liver carries out numerous vital metabolic functions. One of these is detoxification of xenobiotic chemicals. The liver is the first organ that a xenobiotic encounters after absorption from the intestine. As such, it is exposed to higher levels of a toxicant than other organs, which the toxicant reaches after it is more diluted. If the liver is exposed to more of a chemical than it can detoxify, it is adversely affected. Examples of hepatotoxicants (chemicals toxic to the liver) are chloroform, carbon tetrachloride, and other organic solvents. An excess of ethyl alcohol can cause cirrhosis of the liver or can aggravate adverse effects of other chemicals. Excessive alcohol intake is also associated with liver cancer, as are the aflatoxins, which are produced by molds growing on grains and peanuts.

Kidneys

Like the liver, the kidneys can detoxify chemicals. However, their major function is to filter the blood and to eliminate waste products into the urine while retaining water and nutrients such as glucose. In the process of excreting waste substances, kidneys also concentrate them. Thus, they can potentially also concentrate toxic chemicals; that is, they may create the high doses, which then adversely affect them. Examples of nephrotoxicants (chemicals toxic to the kidneys) are some antibiotics and the heavy metals mercury, cadmium, and lead.

Central nervous system

The brain requires high levels of oxygen to function normally. Any substance that lowers available oxygen is neurotoxic (toxic to the brain). Examples are carbon monoxide; certain pesticides, such as malathion and parathion; and nerve gases. Many drugs, legal and illegal, are also neurotoxic.

Reproductive system

Teratogens are chemicals that can kill the embryo or fetus or cause other damaging changes that result in mental retardation, deformed organs, or other birth defects. The embryo (the stage from about the second to the eighth week of pregnancy) is especially sensitive to teratogens. A teratogen is particularly likely to cause damage when the specific organ system that can be adversely affected is developing. The same amount of a teratogen may have no adverse effect if the fetus is exposed after organ formation. Because of this fetal sensitivity to toxicants, pregnant women are advised to avoid use of alcohol, tobacco, and almost all drugs. Even the nutrient vitamin A is a teratogen and must not be taken in excess. The notorious thalidomide is another teratogen. Toxicants or circumstances that adversely affect the mother's health can also harm the embryo or fetus. If the mother is poorly nourished, her malnutrition can adversely affect the fetus. In addition to concerns relating to the embryo or fetus, the germ cells of the mother or father may be affected by a toxicant. In males, the insecticide dibromochloropropane can damage sperm chromosomes and lead to sterility and lead can produce malformed sperm.

Immune system

The immune system is a complex system of cells, tissues, and organs that includes bone marrow, thymus, lymph nodes, and spleen. The immune system recognizes the difference between what is self and what is not-self and works to rid itself of the latter. Examples are microorganisms, other foreign particles, and cancer cells. Immunotoxic substances damage the immune system by either suppressing its function or causing it to overreact. Suppression may damage the immune system's ability to ward off infections or to rid the body of cancer cells. For example, a corticosteroid drug taken over a period of time can suppress the immune system. Drugs capable of suppressing the immune system are deliberately given to people who receive organ transplantation to prevent the body from rejecting the new organ. Some environmental chemicals such as PCBs have been shown to suppress the immune system in animals. Other chemicals or substances such as tree and flower pollens may cause the immune system to overreact, resulting in allergies. The autoimmune diseases lupus erythematosus and rheumatoid arthritis also result from overreactions.

EPIDEMIOLOGIC STUDIES

Epidemiology is the study of the distribution of disease in human populations and the factors that influence the distribution. An epidemiologic study examines potential adverse effects of a chemical at concentrations actually observed within the work place or environment. Historically, epidemiology was successfully used to trace infectious diseases. For example, before water disinfection was practiced, outbreaks of cholera or typhoid fever could be traced to a particular contaminated water supply. It is more difficult to trace a disease to a chemical exposure, but such associations have been successfully made, and associations between disease and people's living and working conditions have been noted for hundreds of years. Several hundred years ago, it was observed that nuns had a higher breast cancer incidence than did married women. In another historic example, in 1775 the English physician Percival Pott observed that scrotal cancer often developed in boys who worked as chimney sweeps. He rightly deduced that their exposure to coal tar was responsible and correctly recommended that the boys should regularly bathe to remove the soot. Before modern safety controls were instituted in the American working place, a number of diseases were traced to occupational exposure to chemicals: Asbestos exposure was linked to the development of asbestosis, lung cancer, and mesothelioma and benzene exposure to blood abnormalities and leukemia. Exposure to high levels of radon in uranium mines was associated with lung cancer.

Epidemiologic studies are most likely to provide useful results when large numbers of individuals are studied, especially if they have been exposed to high concentrations of the suspect substance. A well-known example is the association between cigarette smoking and lung cancer. About 30% of all American cancers are attributed to tobacco use. Likewise, because large numbers of people drink alcohol, it was possible to link excess alcohol consumption with liver cancer. Studies of small populations can sometimes provide useful results if the adverse effect is an unusual one and can be traced to a specific exposure. An instance of this was the linking of vaginal cancers in young women to an exposure they all shared – mothers who took the hormone drug diethylstilbestrol during pregnancy. Another successful linkage between a chemical and disease involved only about 100 individuals who had an uncommon liver cancer – all shared work place exposure to the chemical vinyl chloride.

A *confounding factor* is one that can influence the results in an epidemiologic study independently of the exposure being studied. For example, if a

person uses tobacco or drinks alcohol, this can influence his or her susceptibility to disease quite aside from the exposure being studied. So can a person's
gender or age. It is also difficult to accurately evaluate exposure to the agent
being studied. As a 1995 article in *Science* put it, "Of all the biases that plague
the epidemiologic study of risk factors, the most pernicious is the difficulty of
assessing exposure to a particular risk factor." Information on exposure often
depends on human memories, which may be both faulty and selective. The
way questions are asked can also change the study outcome.

Many epidemiologic studies have been carried out in communities that
have an excessive rate of a particular disease. Almost any community, by
chance, will have an excess rate of some diseases, but people in the community may suspect that the high incidence is due to some shared exposure. For
example, a community's drinking water may contain a particular contaminant, or the community may have a manufacturing facility with chemical emissions of concern. To follow up on community suspicions, epidemiologists look
for links between the exposure that people share and the occurrence of the
disease. Unfortunately, these studies are seldom successful, both because individual exposures are hard to trace and the population is usually too small to
make successful statistical connections. Despite these drawbacks, these studies
continue to be carried out because people suspicious of a facility or waste site
in their community continue to demand them. Because it is so difficult to
reach conclusions in individual communities, investigators are researching a
multisite approach in which results from a number of similar hazardous waste
sites in different locations are pooled. The data are analyzed to see whether
more definitive conclusions can be reached than results for one community
alone. It is too soon to know how successful this approach will be.

Judging the results of epidemiologic studies

A Harvard epidemiologist, Professor D. Trichopoulos, recently observed that
epidemiologists "are fast becoming a nuisance to society. People don't take us
seriously anymore, and when they do take us seriously, we may unintentionally
do more harm than good." Results of epidemiologic studies are commonly reported on television or radio or in the newspaper, sometimes with much fan-
fare. It is very difficult for the person receiving the information to judge how
seriously to take it. However, there are questions that can be asked to help
judge a report:

- Is this the first study of this nature that has been done, or have several studies been carried out, all showing fairly consistent results? Particulates are an
  example of a substance for which a number of studies have consistently
  shown adverse health effects – as ambient air levels of particulates increase,

so do hospital admissions for respiratory diseases and some heart problems. Even with particulates, there are responsible epidemiologists who have differing opinions.

- How many people were studied: many thousands of individuals whose background was well known or only a small community with many possible confounding factors? For example, it was suspected for years that a high-fat diet contributed to breast cancer risk. However, a careful study of 121,000 nurses did not show a link. This study was taken very seriously because it involved a large population whose history was well known and that has been studied since 1976.
- Does the report indicate that the study considered confounding variables, such as age, smoking, diet, or other life-style factors?
- Is the disease a rare one that may be more definitely linked to a risk factor, or is it a very common disease? For example, was there a general increase in cancer or an increase in one specific cancer?
- Did the report indicate how large the risk factor was? A 30% increase in risk or even an apparent doubling of risk may mean little, especially if only one study has been done. However, if there is a three- or fourfold increase in risk, the association carries more weight, even if there is only one study.
- Are study results biologically plausible; that is, can the results be explained in terms of how biologic organisms operate? For example, a number of studies indicate increased death rates in those exposed to environmental levels of particulate matter, but a biologic mechanism to explain this relationship has not been found.
- If clinical work on human beings has been done, are the results of the epidemiologic studies consistent with the clinical results?
- A very important question is, Was the story sensationalized? It is difficult enough to reach conclusions even with carefully presented results. With sensationalized reports it is probably impossible.

The limits of epidemiology

Epidemiology cannot answer all the questions put to it, even if many studies of one particular risk factor are performed, include large numbers of people, and are designed with great care and efforts to control confounding factors. This has been the case with studies of potential adverse effects of electromagnetic fields (EMFs) on humans. Epidemiologic studies, searching for connections between various types of cancer and EMFs, have been ongoing since 1979, but results have not been consistent and are sometimes contradictory. After carefully examining these studies, an NRC panel reported in late 1996 that it had found no evidence of adverse effects on animals or cells at EMF levels found in human residences. One can probably safely say that if there is a

BOX 3.5

Epidemiologic studies remain tremendously useful. Despite problems with and criticisms of epidemiologic studies, they can provide information that no other type of study can. Two recent studies were so highly regarded that they resulted in major recommendations:

1. Earlier epidemiologic studies had associated excess vitamin A intake during pregnancy with increased risk of serious birth defects. A study reported in 1995 strongly supported that conclusion: Pregnant women who had four times the recommended daily intake of vitamin A had a much increased risk of giving birth to babies with cleft lip, cleft palate, and major heart defects. Women of child-bearing age are now urged not to exceed the recommended dose of vitamin A.

2. Another highly regarded study also involved birth defects. In this case, lack of the B vitamin folic acid in pregnant women's diets was strongly associated with serious birth defects: spina bifida (in which the spinal cord is not completely encased in bone) and anencephaly (in which a major part of the brain does not develop). The association was so strong that FDA recommended adding folic acid to grain products to ensure that all women of child-bearing age have a greater intake of this vitamin.

risk from EMFs at levels encountered by ordinary people, it is very small. It is small risks that pose special problems to epidemiologists – there are just too many confounding factors that could influence the outcome of the study. However, very small risks can be important. In the case of EMFs, hundreds of millions of people use electricity and are exposed to EMFs, so even a small risk can translate into an effect on many people. This is the impetus that continues to drive studies of potential EMF risks. Nonetheless, in the future, research work on EMFs will put less emphasis on epidemiologic studies; it will concentrate more on biological studies, looking for effects of EMFs on living cells under carefully specified conditions. Once investigators better understand biologic interactions of EMFs with cells, better-planned epidemiologic studies may be possible.

Remember that even when epidemiologic studies successfully show an association between a chemical and an adverse health effect, that association is not proof. Controlled animal or clinical studies are also needed. Results can sometimes be so overwhelming that they are taken as proof. Consider smoking and disease. Quite aside from the clinical information available, so many associations with disease have been observed with large numbers of people over many years that the results are taken as conclusive.

DISCUSSION QUESTIONS

1. Laboratory studies have shown that about 60 of the estimated 4,000 substances in tobacco smoke are carcinogens, including tar, nicotine, formaldehyde, and benzene. Furthermore, 24 of 30 epidemiologic studies support the conclusion that nonsmokers exposed to secondhand smoke have a greater lung cancer risk than those not exposed. (Secondhand smoke is the side stream smoke emitted between puffs of a cigarette plus the smoke exhaled by the smoker.)
   (a) Do you believe these results indicate that secondhand smoke is always a cancer risk?
   (b) What other information would be useful to know before reaching a conclusion?

2. As compared to urban residents, American farmers have higher rates of several cancers such as non-Hodgkin's lymphoma than do the general population. Environmental factors are suspected of contributing to these rates. Think about the life-style of farmers. In addition to pesticide exposure, what other environmental exposures – chemical or otherwise – might contribute to an increased cancer rate in farmers?

FURTHER READING

Ottoboni, M. A. 1991. *The Dose Makes the Poison: A Plain-Language Guide to Toxicology*, 2nd ed. New York: Van Nostrand Reinhold.

Raloff, J. 1994. Something's Fishy: Marine Epidemics May Signal Environmental Threats to the Immune System. *Science News*, 146(1), 8–9, July 2.

Smith, R. P. 1992. *A Primer of Environmental Toxicology*. Philadelphia: Lea & Febiger.

Taubes, G. 1995. Epidemiology Faces Its Limits. *Science*, 269, 164–9.

4

EVALUATING ENVIRONMENTAL RISK

INTRODUCTION

The word *risk* is defined as the probability of suffering harm from a hazard. A hazard is the source of the risk – it is the substance or situation that can cause harm. A risk can arise from a chemical hazard such as mercury, a biological hazard such as a pathogenic microorganism, or a physical hazard such as ionizing radiation. A risk has a probability ranging from 0 to 1 where 0 indicates that there is no risk at all, whereas 1 represents definite harm. Risk assessment is the systematic examination of the nature and magnitude of risk. What, for example, is the personal risk of being exposed to 3 ppb of the herbicide atrazine in drinking water? Or, what is the risk to a community's health of the air emissions from a local factory? Risk assessment is a process used to help compensate for a lack of information. As a U.S. OSHA official has stated, "People need to understand that we do risk assessment because we don't have conclusive scientific evidence available. Risk assessment is not science; it is a set of decision tools to help us make informed decisions in the absence of definitive scientific information." The tools of risk assessment help us to evaluate specific environmental risks and to set priorities among them, but it seldom provides definitive answers.

This chapter examines chemical risk assessment and comparative risk analysis. As the name implies, chemical risk assessment examines the risk of chemicals. For example, what is the risk to workers in a chemical facility of exposure to benzene? Or, what is the risk to a small child of exposure to household dust containing lead? What is the risk to individuals living near a hazardous waste site of exposure to chemicals at that site? Comparative risk analysis goes beyond examining the risk of individual chemicals, to ask questions such as, What is the risk of stratospheric ozone depletion as compared to that of acid rain? Or, is outside air pollution a greater risk to human health than water pollution?

CHEMICAL RISK ASSESSMENT

Chemical risk assessment follows four steps:

1. Hazard identification: What are the reasons for believing that a particular chemical may be causing harm?
2. Dose-response assessment: Animals are evaluated for their responses to progressively higher doses of the chemical.
3. Exposure assessment: Who is exposed to the substance? What doses are received by exposed individuals? What sources emit the chemical? What is the route of exposure to the chemical? Over what period does exposure occur?
4. Risk characterization: In this final step, information from the first three steps is analyzed and a risk statement is prepared (see Figure 4.1).

Each of these steps is dealt with individually, first for chemicals not suspected of causing cancer and second for potentially carcinogenic chemicals.

Risk assessment for noncarcinogenic chemicals

Hazard identification. Hazard identification is the first step in chemical risk assessment. Remember that any chemical can have adverse effects if the dose is high enough. The chemicals of most concern are those that could be harmful at low doses or at exposure levels that already exist in the environment. Certain chemicals are automatically seen as potential hazards; pesticides are among these because their purpose is to kill. Thus, a new pesticide is tested not only for its effectiveness against pests, but triggers a risk assessment. A new food additive is also a potential hazard as it is intended for ingestion and war-

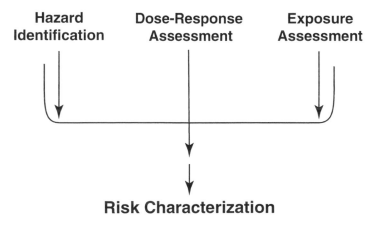

FIGURE 4.1. The chemical risk assessment process.

rants a risk assessment. If results of epidemiologic studies are available, they are also examined when evaluating a chemical hazard.

Dose-response assessment. Once it has been decided that a hazard merits a risk assessment, a controlled dose-response study in animals is the next step. This second step is ordinarily carried out in rats and mice, which are used as surrogates for human beings. In a dose-response assessment, one group, the control group of animals, does not receive the test chemical. A second group is given a low dose on each day of the study over a period of weeks, months, or years. A third group receives a larger dose and a fourth group a yet larger dose. The highest dose that does not exert an adverse effect on the animals is the no observed adverse effect level (NOAEL) (also referred to as the no observed effect level [NOEL]; see Figure 3.1).

The animal data are used to determine a dose safe to humans. A simplified description of the procedure follows: The animal NOAEL is divided by an uncertainty factor, which typically assumes that the average human is 10 times more sensitive to the chemical than test animals. A second assumption is that some humans are 10 times more sensitive than less sensitive humans. This gives an uncertainty factor of 100. If the quality of the dose-response information from animal experiments is not good enough to be sure of the NOAEL, another multiple of 10 is introduced to give a factor of 1,000. Once an uncertainty factor has been selected, the animal NOAEL is divided by that number to arrive at a *reference dose* (RfD), a dose assumed to be safe to humans over the period of exposure; the RfD is also often called the *acceptable daily intake* (ADI). The smaller the value of the RfD, the more toxic the chemical is. Example: What is the RfD of a chemical that has an animal NOAEL of 1 milligram per kilogram of animal body weight per day (1 mg/kg/day)? Assuming the animal test data are of high quality, an uncertainty factor of 100 is used: The NOAEL is divided by 100 to give an RfD of 0.01 mg/kg/day. What would be the RfD of this chemical if the animal data were not of good quality?

Exposure assessment. In the third step of risk assessment, human exposure to the chemical in question is evaluated. Exposure can occur through the route of water, food, air, or soil. Each is examined. If the chemical is found in water, is the water used for drinking? If in food, what foods and how much of each food is eaten? If it is found in air, at what concentrations is it found? In what locations is it found? Special consideration is given to populations that have especially high exposure to a pollutant. For example, PCBs and methylmercury are chemicals that concentrate in fish; this means that people who eat more of the contaminated fish, such as some Native Americans, have higher exposures to these chemicals. Urban dwellers may have higher exposure to motor vehicle exhausts containing carbon monoxide, VOCs, nitrogen

BOX 4.1

Why are factors of 10 often used? In the late 1930s, the first antibiotic drugs, the sulfonamides, came into use. In 1937 a drug company found that sulfonamides could be dissolved in a sweet-tasting organic chemical, diethylene glycol – a chemical closely related to antifreeze. The company marketed sulfonamides dissolved in diethylene glycol without first testing its toxicity in animals. As a result, about 120 people died. This disaster led directly to the 1938 passage of an amended Federal Food Drug and Cosmetics Act (FFDCA) to assure that such a tragedy would not occur again. After discovering the illnesses and deaths that resulted from consuming this chemical, U.S. FDA personnel traveled around the country by train to collect information on the amounts of diethylene glycol that each affected individual had drunk. From this information, they calculated an approximate LD_{50} for human beings and observed that there was about a 10-fold variation in human sensitivity to the chemical. After returning to their laboratories, they tested animal sensitivity to diethylene glycol and found that, like humans, animals varied about 10-fold in sensitivity. FDA then began the practice of dividing the NOAEL by 100, that is, by 10 times 10, to calculate a margin of safety for human exposure to chemicals tested in animals. However, in recent years, as more information on particular chemicals has become available, EPA has begun to move away from routinely using factors of 10.

oxides, and particulates. Small children living in central cities in old deteriorated housing tend to have higher exposure to lead than those living in recently built suburban homes. Urban people who drink chlorinated water are exposed to by-products of chlorine disinfection, whereas rural people who drink well water may be more exposed to radon, nitrate, metals such as manganese or arsenic, and, sometimes, pesticides. Rural dwellers may also have more exposure to airborne pesticides and to farm residues that contain biological toxins. Individuals living near a hazardous waste site may have higher exposures to chemicals found at that site than those living farther away.

The exposure of children is a special case. They breathe more rapidly than adults and, pound for pound, inhale more of an airborne pollutant. Children also have different habits than adults. They may be exposed to soil pollutants by ingesting soil during play. More generally, they ingest things that an adult would not, such as sweet-tasting paint chips containing lead. Pound for pound, they eat more than adults, and, if a pollutant is found in milk or fruits heavily consumed by children, their exposure is greater than that of adults. Children's greater exposure needs to be considered in a risk assessment.

Whether considering adults or children, there is seldom enough information to evaluate exposure adequately. To account for this lack of information and to assure an extra degree of safety, a worst case assumption is made: The degree of exposure is often assumed to be greater than it actually is. An example of the worst case exposure assumption is the procedure used until very recently to evaluate human exposure to pesticide residues on food. This assumed that a given pesticide was used on all crops on which it was legal to use it, at the maximum level at which it was legal to use it, and also assumed that the crops were harvested at the shortest allowable interval after pesticide treatment. On the basis of such assumptions, EPA halted many uses of the EBDC family of fungicides in 1989. Later, after a $10 million study of actual residues in food sold to consumers, it was found that 80% of the foods had no EBDC residues at all. On the basis of these results, EPA again allowed EBDC fungicides to be used for many purposes. As this example shows, exposure can be more reliably estimated if the assessment has sufficient input of money, time, and other resources.

Risk characterization. Risk characterization is the fourth step of the risk assessment process. To do a risk characterization, all the information collected from the hazard identification, dose-response, and exposure studies is ana-

---

BOX 4.2

Cancer. The word *cancer* is used to describe abnormal cell growth – cells multiplying without the usual controls. Abnormal cell growth may develop after mutations in a cell's genetic material, its deoxyribonucleic acid (DNA) or genome. For cancer to develop, as many as 10 distinct mutations may first need to accumulate in the DNA. As one cancer researcher expressed it, "In cancer the genome is shot to hell." In the United States, incidences of some cancers have increased and those of others have decreased; for example, breast and skin cancer rates have gone up, but stomach cancer rates have decreased. The risk of cancer increases as a person ages. By age 85, cancer will have developed in about a third of people. Thus, the increasing average age of the American population has led to an increased number of cancers. About one in three Americans is diagnosed with cancer in his or her lifetime, and about one in five will die.

Life-style plays an important role in cancer development. About a third of all cancers are associated with tobacco use. Diet is also tremendously important. Ingesting too much fat in the diet or too little fiber is associated with an increased risk of colon cancer and possibly an increased risk of skin and prostate cancer. Obesity increases cancer risk.

BOX 4.2 *(continued)*

Long-term high-level consumption of salted and pickled foods is associated with stomach cancer. Heavy alcohol consumption is associated with an increased risk of liver, colon, and breast cancer. Sexual habits affect cancer risk; for instance, the human papilloma virus (HPV), which is associated with cervical cancer, is more common in women who have had multiple sexual partners – and even more common in married women whose husbands have had multiple partners. Exposure to ultraviolet (UV) light through excessive sun exposure heightens the risk of skin cancer.

Other factors that affect the development of cancer include medical and dental X-rays, which are a risk factor for leukemia. Viral and bacterial infections are associated with some cancers, as HPV is associated with cervical cancer. Infection with the hepatitis B virus enhances liver cancer risk; in some parts of Asia and Africa where hepatitis B infection is common, liver cancer is one of the five most common cancers. Worldwide, this virus is one of the most important carcinogens, second only to tobacco, but it is much less common in developed countries. Recent studies indicate that infection with the bacterium *Heliobacter pylori* heightens the risk of stomach cancer. Occupational and environmental pollutants are believed to cause some cancers; one study, which made a worst case assessment of the cancer risk of all known environmental pollutants that are carcinogens, came to the conclusion that 1% to 3% of cancers may be due to environmental pollution. Studies by epidemiologists have arrived at a similar figure.

How can a chemical in tobacco smoke, a mold toxin, or some other chemical cause cancer? Some carcinogens are initiators; that is, they directly affect DNA to cause a mutation. Theoretically, a cancer initiator does not have a threshold – a safe dose – and any dose greater than zero is assumed to pose some risk. Other chemicals are promoters; that is, they promote cancer, but not by directly affecting DNA. A cancer promoter does have a threshold, below which it does not promote cancer. Some carcinogens both initiate and promote cancer; these are called the complete carcinogens. It is important to point out that damage to DNA does not mean that cancer will occur. Animals have evolved mechanisms to repair damaged DNA. As two Canadian environmental health workers noted in the August 1995 issue of *Environmental Science & Technology*, "Cancer occurrence may be related more to the failure of DNA repair capability than to the degree of trace exposures to DNA damaging contaminants."

lyzed. The risk is then calculated and the information given to risk managers, who are responsible for finding means to reduce the risk.

Risk assessment for carcinogenic chemicals

Hazard identification. Hazard identification is the first step taken to assess chemicals suspected of causing cancer. However, only a limited number of potential carcinogens can be studied because cancer risk assessments are lengthier and costlier than are those for chemicals not suspected of causing cancer. Each may take 4 to 6 years to complete and cost several million dollars. This makes the hazard identification step used to choose the chemicals to be tested especially important. What reasons are there to suspect that a particular chemical is a carcinogen? Is epidemiologic information available that makes the chemical suspect? What if there is good reason to suspect that a particular chemical is a carcinogen, but only a handful of research personnel are exposed to it? In this case, a cancer risk assessment probably will not be made; it may be sufficient to warn individuals who use the chemical to take proper precautions. But if the suspect chemical is produced in large amounts and many people are exposed to it, a costly long-term study can be justified. In an earlier era, when safety controls in the American work place were often inadequate, evidence that a chemical was a carcinogen was sometimes found in the work site, for example, occupational exposure to benzene, a widely used industrial chemical, was found to be associated with a heightened risk of leukemia. Only later was benzene's ability to act as carcinogen confirmed by animal studies. Vinyl chloride and asbestos are other chemicals whose hazards were first observed through workplace cancers. However, the intent in modern-day America is to deduce a chemical's hazards before it adversely affects people and then to control exposure to it. Even better, the P^2 ethic urges using less of the chemical or, in some cases, discontinuing its use.

Dose-response studies. The National Toxicology Program (NTP) recommends the following protocol to test chemicals that may cause cancer: Two animal species, rats and mice, are typically tested. Starting after weaning, the chemical is administered to the animals every day for 18 months to 2 years (lifetime studies). If possible, the chemical is added to the diet. If animals will not accept it in their food, it is given through a stomach tube, a process called *gavage*. For airborne chemicals, animals are exposed to the chemical in an enclosed chamber 5 days a week for 6 hours each day. The doses administered to animals are calculated as milligrams per kilogram (mg/kg), where the daily dose is measured as a ratio of milligrams given per kilogram body weight. Each dosage is tested in two groups, one of 50 males, another of 50 females. A first group, the controls, receives no chemical. A second group receives the maximum tolerated dose (MTD) of the chemical. A third test group receives one-half the MTD. A fourth group, which receives a lower dose, is often included. The MTD is the highest dose that does not reduce survival of the animals as a result of causes other than cancer. It is determined from studies in which the animals are exposed to varying doses of the chemical for several months.

If the chemical gives rise to excess tumors – more than expected as compared to the control group receiving no chemical – in these lifelong studies, its cancer risk is calculated. The relative cancer risk of some chemicals is given in Table 4.1. As seen in the table, there are huge differences in the potency of

TABLE 4.1 Relative cancer risks of selected chemicals

CHEMICAL	CANCER SLOPE FACTOR $(mg/kg\text{-}day)^{-1}$
2,3,7,8-TCDD	100,000
Aflatoxin B_1	2,900
Ethylene dibromide	41
Arsenic	15
Benzo[a]pyrene	5.8
Cadmium	6.1
PCBs	4.34
Nickel	1.05
DDT	0.34
Chloroform	0.081
Benzene	0.029
Methylene chloride	0.014

Note: Values given reflect relative cancer risk. As examples: 2,3,7,8-TCDD is more than 30 times stronger as a carcinogen than aflatoxin B_1, which, in turn, is 8,500 times stronger than DDT.

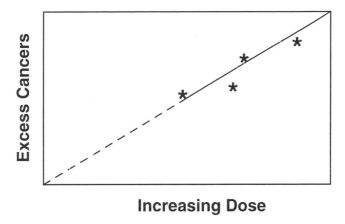

FIGURE 4.2. Excess cancers and increasing dose of a carcinogen.

carcinogens, that is, in their ability to cause cancer. The most potent rodent carcinogen is 10-million fold stronger than the weakest. The higher the value shown in Table 4.1, the more potent is the carcinogen. Among the chemicals shown, 2,3,7,8-TCDD is the most potent. The mold toxin aflatoxin is more than 30-fold less potent. Compared to TCDD or aflatoxin, the other chemicals shown are weak carcinogens. To calculate human cancer risk using this information, two important assumptions are made: It is assumed that if the chemical causes cancer in animals, it can also cause cancer in humans. The second assumption is that it is possible to extrapolate from the high doses used in animal studies to low-dose human exposures. In Figure 4.2, the portion of the line drawn through the stars represents the "best fit" relationship between the doses tested in animals and the response (excess cancers) observed at each dose. For doses lower than those tested in animals, the line is continued as a dashed line down to a dose of zero; this represents the policy assumption that any dose greater than zero poses some risk.

In Figure 4.2, as noted, the lower portion of the line represents an assumption that there is no safe dose of a carcinogen, that even one molecule of a carcinogen poses a finite, although theoretical risk. This is an example of a worst case or default assumption. It is an important assumption because people are ordinarily exposed to doses that are hundreds, thousands, or hundreds-of-thousands of times lower than doses tested in animals. Worst case assumptions have been used by U.S. regulatory agencies as a means of erring on the safe side. As a former FDA administrator expressed it, "When science fails to provide solutions, FDA applies conservative assumptions to ensure that its decisions will not adversely affect the public health." Another example of a worst case assumption is assuming that all carcinogens are initiators; that is, they are assumed to affect DNA, the genetic material, adversely. Other worst case assumptions include considering a chemical a carcinogen even if it causes tumors only at MTD, in only one sex, one species, or one strain of a species. In fact, strains that are prone to development of cancer are often deliberately used. For example, the chemical trichloroethylene was classified as a carcinogen because, when it was given at MTD to a strain of mice prone to development of cancer, the males showed excess tumors.

However, even cancer initiators follow the dose-response rule of toxicology: As the dose increases, the risk increases. England, Denmark and the Netherlands take a different approach to carcinogens than the United States. They believe that carcinogens do have safe doses (thresholds). This means the dose-response curve would not be straight down to zero as seen in Figure 4.2, but would have the shape observed in Figure 3.1.

Risk characterization. The information from hazard assessment, dose-response assessment, and exposure assessment studies is analyzed and a risk for the chemical is calculated. The dose-response studies provide a cancer potency factor (Table 4.1 has examples), and the exposure assessment provides an average intake of the chemical over a stated period (often 70 years). Multiplying the cancer potency by the average intake gives a rough estimate of the relative risk.

Once the cancer potency of a chemical and exposure to it is known, the risk can be calculated. The risk is often expressed as the increased chance of development of cancer over a 70-year lifetime. In each of the following examples, it is assumed that for 70 years a 70-kg person drinks 2 liters of water containing the chemical at the concentration noted per day:

- Radon at 300 pCi/L: Cancer risk is increased by 2 in 10,000 (that is, in a population of 10,000 people, 2 additional cases of cancer are projected).

BOX 4.3

Caveats. The following caveat has been routinely added to cancer risk assessment documents by regulatory agencies: "These estimates represent an upper bound of the plausible risk and are not likely to underestimate the risk. The actual risk may be lower, and in some cases, zero." Also remember that the number calculated from a risk assessment is a theoretical quantity. Consider the risks of driving a car. We know that about 40,000 Americans are killed each year in driving accidents so we can accurately calculate the known risk of death that a person faces when driving. However, the numbers obtained from chemical risk assessments are theoretical – no one really knows the exact risk.

- Arsenic at 2 µg/L: Cancer risk is increased by 1 in 100,000.
- Tetrachloroethylene at 2.4 µg/L: Cancer risk is increased by 1 in a million.

The regulatory goal is a negligible risk, also referred to as a virtually safe dose or a de minimis risk. An excess cancer risk of one in a million is often considered a negligible risk. Dr. Frank Young, a former FDA commissioner, has stated, "When FDA uses the risk level of one in a million, it is confident that the risk to humans is virtually nonexistent." An excess risk no greater than 1 in 1 million is the goal for known human carcinogens. Some consider a risk of 1 in 100,000 acceptable for chemicals known to be carcinogens only in animals.

CHANGING THE RISK ASSESSMENT PROCESS

Problems with risk assessment

In the past 10 years, there has been increasing criticism of the cancer risk assessment process, especially of the use of default (worst case) assumptions. A major argument expressed is that the MTD is toxic and leads to rapid cell proliferation. During rapid proliferation, the genetic material replicates rapidly, a condition that increases the likelihood of mutations that could cause cancer. In other words, rapid cell proliferation may be the actual carcinogen. Indeed, evidence is accumulating that high doses of chemicals can stimulate cell proliferation. The reason a possible carcinogen is tested at very high doses is because it is often impossible to test at low doses. This is true because in at least 30% of the control rodents – animals not given the chemical – cancer develops by the end of their lives. Thus, unless the chemical gives rise to an unusual cancer not seen in control animals, toxicologists must look for excess cancers. This process of administering high doses and then looking for excess cancers has been called a black-box approach because no one knows how the excess cancers were produced. Recent studies have increasingly examined the mechanism (the biologic mode) by which a carcinogen causes cancer. If the mechanism is known for a chemical, the black-box approach is not needed. It becomes possible to know when it is legitimate to extrapolate from high animal doses to the lower doses to which humans are exposed and, conversely, to know when it is reasonable to assume that low doses will not harm humans.

The following examples are among those used by those who criticize the use of worst case assumptions. Use of the pesticide dichlorvos in pet flea collars was halted in the late 1980s although 10 published studies had concluded that dichlorvos was not a carcinogen and Japanese studies independently concluded that it was not a human carcinogen. Only one study showed excess cancers in rats. Even that study was criticized by an EPA science advisory panel as having serious flaws, but dichlorvos was nonetheless reclassified as a possible carcinogen. Another example was the proposed banning from foods of red dye number 3 by the FDA. This dye, which gives cherries in fruit cocktail their characteristic color, was shown to cause excess thyroid cancers when fed to rats at a level of 4% of the diet. For a human to ingest a comparable amount of red dye number 3, that person would have to eat 13,898 servings of fruit cocktail every day for 70 years. Nonetheless, until the Delaney Clause was repealed recently, red dye number 3 was slated to be banned from fruit cocktail. Another example of worst case assumptions is very important because it affects drinking water that is disinfected by using chlorine. The chemical chloroform is formed during chlorine disinfection. Chloroform was deemed a carcinogen after large doses were daily injected into the stomachs of mice and rats and more liver cancers later developed in these animals than in controls. On the basis of these results, EPA developed a drinking water standard for chloroform of 0.004 ppm. However, it was later shown that mice exposed to chloroform in their drinking water at levels up to 1,800 ppm – a concentration 450,000 times greater than the 0.004 ppm standard – did not show liver or kidney toxicity. At levels of chloroform similar to those found in drinking water, the mice showed no increased cancer risk. EPA scientists acknowledge that chloroform's risks were inadvertently exaggerated.

Critics believe that overuse of worst case assumptions leads to a loss of what could be valuable chemicals and pharmaceuticals. Even if a chemical categorized as a carcinogen is used, stringent standards make its application extremely expensive. At the same time, a different chemical that is not a carcinogen may be considered a better choice even if it presents other hazards. However, worst case chemical risk assessment has staunch supporters who point out that it allows truly important problems to be identified and prioritized. They believe that it is not the risk assessment process itself that should be criticized, but the misuse of risk assessment results by those who misrepresent them to support their own aims. As former EPA administrator Ruckelshaus noted, "Risk assessment is like a captured spy. If you torture it long enough, you can get it to tell you anything you want."

In 1990, the CAA Amendments required EPA to ask the NAS to review the risk assessment process and make recommendations for improvement. In 1994 the result of the NAS deliberations, *Science and Judgment in Risk Assessment*, was published. The report analyzed current approaches to risk assessment and endorsed EPA's overall approach, but suggested many ways to improve the process. The NAS panel evaluated EPA's use of default or worst case assumptions, which some members believed should be used only as a last resort, in instances in which so little information is known that assessors have no basis on which to base expert judgments. However, default assumptions cannot be entirely avoided because dose-response studies and other means of analyzing how a chemical acts in an animal body seldom produce definitive results. Likewise, exposure assessments seldom provide all the information needed. Because we don't have all the information needed, a case can be made for cautious assumptions. One important request that the NAS made was that EPA should provide risk managers with more information than it had in the past: a narrative describing the risk and a range of numbers for the risk, not just one.

New guidelines for risk assessment

In response to NAS recommendations, EPA developed new guidelines, officially proposed in 1996, to modernize cancer risk assessment. In the past, the focus was simply, How many excess cancers does this chemical produce? In the future, additional important information on the chemical will be considered. Most importantly, the question, How does the chemical cause cancer?, will be asked. If the evidence indicates that high doses of the chemical cause cancer by stimulating cell proliferation, that information can be considered. So can data on dose and response and route of exposure to the chemical. Being able to consider all available information will allow risk assessors to evaluate better whether the chemical is likely to act in the same way in humans as in animals, and to determine whether human exposure to low doses of the chemical poses a risk. It will no longer be automatically assumed, as in the past, that a carcinogen is dangerous at any dose greater than zero. Rather, if animal studies indicate that it has a threshold, this will be considered. Populations that are more susceptible to the contaminant will also be identified; children are often one such population. More attention will be focused on chemicals that actually are most likely to affect people adversely, and less on chemicals that cause cancer in animals only at very high doses. In the past, after risk assessors classified a chemical as a carcinogen, they calculated a number indicating the degree of risk of that chemical, which they passed on to the risk managers. In the future, the assessors will prepare a narrative for the managers, one that lays out key evidence as to whether or not the chemical is a carcinogen, what its mode of action is, and under which conditions it is a hazard. The narrative will also note the strengths and weaknesses of the data.

Risk management

Scientists carry out risk assessments. They give the information they obtain to risk managers – the individuals who decide how to lower the risk. Risk managers may be legislators or they may be regulators in agencies like the EPA, FDA, and OSHA. The major factor driving risk management is risk reduction, but other factors enter into decisions. These include the statutory requirements, that is, what the law requires and the technologies available to reduce the risk. For example, what technologies are available to capture a pollutant end-of-pipe? What are the public's concerns? What are the political concerns? What does it cost to control the risk? Risk managers have other tools in addition to regulation: Educating the public about the risk and what they can do about it is one. Requesting that industry voluntarily comply with certain guidelines in order to avoid regulation is another. Cost has become a key issue.

The new risk assessment guidelines discussed are expected to change how risk will be managed in the future greatly. Risk managers will find their task more difficult because, as one government official put it, risk managers "will

---

BOX 4.4

Mixtures. Chemical risk assessment analyzes one chemical at a time, but it is not enough to consider only individual chemicals. Humans, animals, and plants are exposed to mixtures containing many chemicals, such as those found in gasoline vapor, urban air, or drinking water. Fish and other aquatic organisms living in a stream receiving discharges from an industrial or municipal facility are exposed to complex chemical mixtures. Industries that discharge wastewater into a receiving stream are now required to examine *whole effluent toxicity*, that is, the effect of the whole effluent. Fish downstream of certain Canadian pulp and paper mills, for example, have shown impaired reproductive development and reduced fertility. But, it is not known which effluent chemicals are responsible. Fish living downstream of English municipal wastewater treatment plants are similarly impaired, but, to date, the chemicals responsible have not been identified.

The situation becomes even more complicated when combined effluents from a number of sources show adverse effects. Otters living in Oregon's Columbia River have impaired reproductive development as compared to those living in less polluted water. There are many contaminants – dioxins, pesticides, heavy metals, or a combination of these – that could be responsible for the adverse effects. Not only are the chemicals responsible for these effects unknown, but, because many sources contribute to the pollution, it is impossible to pinpoint blame. Although determining whole effluent toxicity is increasingly important, it does not negate the need for doing risk assessments on individual chemicals. Consider a hazardous waste site: Although it typically contains many chemicals, only a few may pose the lion's share of the risk and it is important to consider these individually.

really have to understand the decisions they are making." Over time, it is expected that the new guidelines will profoundly affect how the federal government regulates pesticide use, water discharges and drinking water treatment, industrial air emissions, and standards for cleaning up hazardous waste sites. In some cases, regulations may be relaxed, but in others they may be tightened.

DISCUSSION QUESTIONS

1. Of the approximately 300 chemicals found in roasted coffee, 11 have been tested for their ability to cause cancer in mice. Eight of the 11 increased cancer incidence. One cup of coffee contains about 10 mg of these 8 chemicals, the equivalent of several months' intake of all synthetic carcinogens combined. Epidemiologic studies have shown no increased cancer risk associated with drinking coffee.
   (a) Would you limit your coffee intake on the basis of this information? Explain.
   (b) On the basis of information you have now been given, would you be more concerned about carcinogens in coffee or those in cigarette smoke? Explain.

2. Consider the following examples: (1) The action level for radon in U.S. homes, that is, the level at which EPA recommends that action be taken to reduce it, is 4 pCi/L. (Picocurie [pCi] is a unit of radioactivity.) In Canada, the action level is 20 pCi/L. (2) In the United States, the human dioxin intake that is considered virtually safe is 0.006 pg/kg body weight/day, whereas in Canada it is 1 pg/kg/day, or 167 times greater. Canada has access to the same information as the United States.
   (a) What are possible reasons that policy makers in different countries may come to different conclusions?
   (b) Is it always good policy to play it safe and impose more stringent regulations? Defend a yes answer to this question. Then defend a no answer.

3. A hazardous waste site has been discovered near your community. Representatives of state health and environmental protection agencies are visiting to discuss implications of the site with concerned citizens and to answer questions. What questions will you ask and why are those questions important to you?

4. A local municipal solid waste combustor recently malfunctioned and showered ash over your community. Some children and elderly people living near the facility complained of respiratory problems at the time of the incident.

    (a) What questions will you ask representatives of health and regulatory agencies about this incident?

    (b) Would you press for a complete risk assessment on the basis of this incident? Give a background for your opinion.

    (c) Would the questions you ask and the demands you make be the same if the malfunctioning facility were a hazardous waste incinerator?

5. (a) Assume that a specific chemical is a carcinogen only at or near MTD. In your opinion, is it justifiable to treat any dose of that chemical greater than zero as posing a cancer risk? Explain.

    (b) Recall that the Delaney Clause mandates zero risk for synthetic carcinogens in processed foods. A person wrote a letter to *Environmental Health Perspectives* defending Delaney and stating that it should be expanded, not eliminated. The reason he gave was that any alternative approach would have to define some level of "acceptable risk," which he believed would necessarily provide less protection of public health. Do you agree or disagree with his position? Why?

    (c) Western European countries use methods to deal with chemicals that do not pose a cancer risk that are similar to those of the United States, but they treat carcinogens differently. If Western European countries are especially concerned about a carcinogen, they may try to ban it. Otherwise, they treat it as a noncarcinogen, that is, as having a threshold and thus an acceptable level of exposure. Criticize this approach to handling carcinogens. Now, develop arguments to support this approach.

COMPARATIVE RISK ANALYSIS

For years, American environmental policy was set piecemeal. Congress reacted to public concerns by enacting laws telling EPA to regulate specific problems. The 1970 Clean Air Act was passed to deal with air pollution, the 1972 Clean Water Act to deal with water pollution, and the 1974 Safe Drinking Water Act to deal with drinking water pollution. These and other environmental laws resulted in about 70,000 pages in the Code of Federal Regulations. These command-and-control regulations became increasingly complex over the years and sometimes redundant and even contradictory. For businesses and municipalities who must comply with the regulations, they mean carrying out a tremendous amount of monitoring and control, analysis, and paperwork. There are limitations to command-and-control regulations and society has limited resources. By the mid-1980s, it was clear that this approach to environmental protection was inadequate and sometimes counterproductive. Large amounts of money were spent on certain problems while others, equally or more serious, received limited funding. The possibility that certain problems were being overlooked entirely was disturbing. The need to deal with environmental issues in their totality, rather than individually, was also recognized. One alternative to command-and-control, discussed in Chapter 2, is P². However, P² is also not the total answer. In 1986, EPA managers began to ask, Which environmental issues pose the greatest risk? How well do the nation's dollar expenditures correspond to actual risks?

William Reilly moved the comparative risk process forward when he became EPA administrator in 1989. He wanted to focus attention on the big risks even when those risks were not receiving the most public attention. In a 1990 speech reported in *Science News*, Reilly compared our approach to environmental problems to the video game *Space Invaders*: "Every time we saw a blip on the radar screen, we unleashed an arsenal of control measures to eliminate it. In Space Invaders, the yardsticks of success are time and casualties, not bang for the buck. Because gunners never run out of ammunition, they have little incentive to take careful aim before blasting at every enemy in sight with both barrels. But in the real world of budget deficits, trade deficits, tax revolts and recession rumors, ammunition is a metered commodity. The time has come to start taking aim before we open fire." Administrator Reilly asked EPA's Science Advisory Board (SAB), a group of 39 nationally recognized scientists and engineers, to make a comparative risk analysis, that is, to rank environmental risks on the basis of their relative seriousness and to suggest ways to reduce risks.

Comparative risk analysis has the goal of comparing the relative seriousness of a number of environmental risks. As such, it is a much broader effort than is chemical risk assessment, but its steps are analogous. As with chemical risk assessment, the first step is to identify risks, in this case, the risks that are to be compared. The second step, analogous to dose-response and exposure assessments, is to analyze the risks identified. However, no new research is ordinarily carried out. Instead, the analysis depends upon information already available. The final step is a risk ranking to compare the severity of risks, one to another.

Table 4.2 notes criteria with which to analyze identified risks to human and ecological health. The problem may already be present and already threatening or continuing to threaten human or environmental health; lead in the environment and its continuing risk to small children are examples. Another is acid rain with its continuing risk to water bodies and forests. Or the risk may be theoretical, but important to consider, as is the case with some chemical emissions. Or, like global warming, the risk is still largely in the future; however, because its impact could be so great, it must be considered now. Other criteria that need to be considered are noted in the table. Other questions

TABLE 4.2 Analyzing environmental risks to human and ecological health

CRITERION AND EXPLANATION	EXAMPLES
Scope of effect	
1. How large a space is exposed?	Region of a state, the nation, world.
2. How many people are exposed?	10,000 people living near a waste site. Millions exposed to ground level ozone.
3. How much of resource is exposed?	One portion of forest, whole forest, or many forests.
Likelihood of an adverse effect	
1. Among people exposed to risk?	Cancer risk of one in a million over 70 years of exposure to chemical in drinking water. Millions immediately at risk from pathogen in drinking water.
2. In resource exposed to risk?	Acid rain falling on alkaline soil that can neutralize it. Falling on acidic soil that cannot neutralize it.
Severity of effect if it occurs	
1. Among people exposed to risk?	Few thousand with temporary skin rash. Lowered IQs in millions of children.
2. In resource exposed to risk?	Fish growth in one lake stunted for one season. Millions killed due to chemical spill in river.
Trend for pollutant	
1. Is its environmental level increasing, decreasing or same?	Metal levels increasing. PCBs and dioxins decreasing very slowly.
2. What is its life span in the environment?	Ground level ozone (hours to days). Methane (10 years). Metals (possibly centuries).
3. Does it bioaccumulate?	Polychlorinated chemicals. Methylmercury.
Trend for other environmental stresses	
1. Is amount of resource or its quality remaining the same or being degraded?	Wetlands being lost, degraded. Habitats being lost, degraded.
Time necessary to reverse adverse effect if it does occur?	5-20 years to recover from oil spill. 150 years to remove excess carbon dioxide from the atmosphere.

must also be asked. One major additional question is, What is the quality of information on the risk? Often, there is too little information, or the information may be too inconclusive to permit firm conclusions or, sometimes, even tentative conclusions.

The SAB report

In late 1990 the SAB produced a report on its deliberations called *Reducing Risk: Setting Priorities and Strategies for Environmental Protection*. As shown in Table 4.3, the SAB Ecology and Welfare Subcommittee ranked risks to the environment as high, medium, or low. However, the Human Health Subcommittee did not believe it had enough information to rank risks to human health. Instead it selected four relatively high-risk problems, but did not rank them in relation one to each other. SAB also made a number of recommendations to EPA:

• EPA should target its efforts to maximize risk reduction. High-risk problems should receive more attention in EPA's planning and budgeting process.

TABLE 4.3 Risks of environmental problems as ranked by EPA's Science Advisory Board

RISKS TO HUMAN HEALTH

Ambient (ground level) air pollutants
Worker exposure to chemicals in industry and agriculture
Indoor air pollution
Drinking water pollution (chemical and microbial)

RISKS TO NATURAL ECOLOGY AND HUMAN WELFARE

Relatively High-Risk Problems

Habitat alteration and destruction
Species extinction and loss of biodiversity
Stratospheric ozone depletion
Global climate change

Relatively Medium-Risk Problems
Herbicides and pesticides
Acid deposition
Airborne toxics
Toxics, nutrients, biochemical oxygen demand, and turbidity in surface waters

Relatively Low-Risk Problems
Oil spills
Ground water pollution
Radioactive chemicals
Acid runoff to surface waters
Thermal pollution

Note: The human health risks are all considered relatively high risks and deliberately not ranked in any particular order. Risks to natural ecology and human welfare are ranked high, medium, or low.

- EPA should spend its discretionary funds on high risk problems. Unfortunately, those funds are very small because Congress limits what EPA can do with appropriated funds. As a long-range solution to this, SAB advised EPA to work to educate both Congress and the public about the need to set priorities.

- EPA should minimize the use of command-and-control regulations and emphasize other tools to reduce risk. Providing information to industry and to the public is one such tool. Remember the Toxic Release Inventory described in Chapter 1. Businesses had only to inform the public of the emissions of these chemicals – no other action was required. But the business community was uncomfortable with the information it was required to report and began to examine means to limit emissions. At the same time, a newly informed public began to exert pressure on them to reduce emissions. Another tool that goes beyond command and control is use of market incentives. Sulfur dioxide trading by electric utility plants is the major market incentive now operating. Each utility has a sulfur dioxide emission allowance. If it emits less than its allowance, it can sell the unused portion to another utility that is emitting more than its allowance. This allowance trading is criticized by some as a license to pollute. Others see it as a means to give an electric power plant the flexibility to make the long-term plans most fitting to its particular circumstances. Another example of a market incentive is consumer deposits on products like beverage containers, tires, or furniture, which may later be refunded.

- EPA should attach as much importance to reducing ecological risk as it does to reducing human health risks. EPA now spends more than 90% of its money on human health protection. However, the report stressed the close link between human health and the health of wetlands, forests, oceans, and estuaries and noted, "This recommendation is not meant to imply the relative value of human life *vis a vis* plant or animal life. Rather, it is meant to reflect in national environmental policy the very strong ties between all forms of life on this planet: Ecological systems like the atmosphere, oceans, and wetlands have a limited capacity for absorbing the environmental degradation caused by human activities. After that capacity is exceeded, it is only a matter of time before those ecosystems begin to deteriorate and human health and welfare begin to suffer."

    Ground level ozone is a risk to human health, and an example that demonstrates the relationship between ecological risk and human health risk. Many trees, crops, and other vegetation, as will be discussed in Chapter 5, are even more sensitive to ozone's adverse effects than humans, and humans directly depend on the plant world for existence. So should the ozone standard be set at a lower level than is necessary to protect human health so that vegetation is protected?

Another example is agricultural land. Agricultural land is not part of our natural environment, but environmental impacts on agricultural land directly affect human welfare. Consider that since 1955, nearly a third of the world's agricultural land has been abandoned because of damage such as soil erosion or nutrient depletion caused by livestock overgrazing and various harmful agricultural practices, including not allowing fallow periods between crops. This is a land area greater than that of India and China combined. People who previously farmed land that is abandoned may be forced into other environmentally undesirable practices. They may move onto land not previously farmed, such as steep hillsides, which then become very susceptible to erosion, or contribute to deforestation by cutting trees to provide agricultural land. At the same time, in the United States as well as in poor countries, more and more agricultural land is being covered with homes, businesses, and roads. This loss of agricultural land is happening at the same time that human population is rapidly increasing.

- EPA should promote the development of methods that will allow society to place a value on natural resources: What is the value of clean water? Clean air? A healthy forest? A sustainable supply of fish? Economic methods have traditionally undervalued these important ecological resources despite the fact humans depend on them for life itself.
- EPA should work to improve public understanding of environmental risks and work to reduce the gap between how the public understands risk and how scientists understand it. For example, the public believes that hazardous waste sites are a major health risk, a view EPA's Advisory Board did not share. It is Congress, often with significant public pressure, that makes laws, appropriates funds to EPA, and specifies how EPA must use the funds. Regardless of scientific opinion, a significant portion of EPA's budget has been going toward hazardous waste site cleanup. Furthermore, EPA is a regulatory agency, which takes laws developed by Congress and develops regulations based on those laws. EPA often has neither time nor money to take other actions, even very sensible actions.
- EPA should include broad public participation in its efforts to rank environmental risks. Public values are an important factor that scientists may be unable to rank adequately. The environmental problem may involve values that the public cares deeply about like aesthetics, peace of mind, and fairness in how the risk is shared, how future generations may be affected, and how the economy may be affected.
- EPA should emphasize P^2 as the preferred means to reduce risk. The SAB report stated "The costs of cleaning up and disposing of pollutants after they have been generated can be enormous. End-of-pipe controls and waste disposal should be the last line of environmental defense, not the front line."

Scientific uncertainty. Given exactly the same information, different scientists may reach different conclusions. This scientific uncertainty makes risk ranking, an already difficult job, even more difficult. One subject evaluated by the SAB is global warming; the primary uncertainty is whether a serious amount of warming will occur. Global warming has serious implications: The whole earth would be affected, potential impacts are severe, and it would take at least a hundred years for accumulated greenhouse gases to dissipate. In this case, despite uncertainty, global warming was placed in the high-risk category. Another example of a disagreement centers around hazardous waste sites, which the SAB did not see as a priority risk. Others, including scientists at ATSDR, whose job is to assess hazardous waste sites, take them more seriously; they point out that 11 million Americans live within 1 mile of seriously polluted sites and that some epidemiologic studies indicate that site contaminants may adversely affect the health of those living nearby. Even if the SAB agreed with this more pessimistic viewpoint, it is not clear that it would rank hazardous waste sites as a high risk *compared* to other risks. Remember, the key word is *comparative*.

For any risk, as more information becomes available, scientific conclusions evolve, and beliefs about the relative seriousness of a risk may also change. Cancer is an example. As our understanding of cancer development improves, fewer scientists are concerned that very low levels of exposure to environmental chemicals will cause cancer. On the other hand, as more is learned about particulate matter air pollution, concern has heightened that it is seriously aggravating respiratory and other health problems.

Differing perspectives on risk. The fact that the SAB ranked a specific problem as low risk does not mean that it should not be taken seriously. It rated oil spills as relatively low risk, but the local effects of oil spills can be devastating, as demonstrated by the effects of the *Exxon Valdez* spill on birds, mammals, and other wildlife. Or, consider that 8 years after an oil spill along Panama's Caribbean coast, a Smithsonian Tropical Research Institute study found that mangrove trees, coral reefs, and the plant and animal life they support still had not recovered. These points are made not to dispute SAB findings, but to emphasize that, although effects may be local, may not affect a large geographic region, and may not persist as long as a phenomenon like global warming, they can be tragic. In other words, low risk does not mean that action should not be taken. However, on the national level, the problem probably warrants fewer resources than one affecting a larger geographic locale or one with a recovery time that reaches into decades and even centuries.

The risk assessment process is important not only at the national level, as analyzed by the SAB, but at regional, state, and local levels as well. Local conditions vary and specific areas need to determine what risks are most significant to them. The SAB did not see hazardous waste sites as a priority hazard, whereas a state with many hazardous waste sites might rank them higher. The SAB also saw groundwater pollution as a relatively low-risk problem, but agricultural states using large amounts of pesticides and fertilizers may rank it higher. There are many risks that may be more important to one region or state than to the United States as a whole. Since the SAB study was completed in 1990, EPA has carried out several regional risk comparison studies. Many states have also done comparative risk studies. In the same way that a specific region or state may have priorities that differ from national priorities, a specific community within a given state may have its own priorities. For example, Detroit has been devastated by a large number of deserted factories and businesses, most left in bad condition. These impair the ability of the city to function. Or, consider Roxbury, a Boston inner city community. It has 80 auto repair shops in a 1.5- square-mile area; trash transfer stations; heavy-duty truck traffic, often idling as the trucks wait to take on or drop off materials; food processing facilities with bad odors; furniture stripping sites; pervasive dust from many points; a number of hazardous waste sites; and a large number of houses with leaded paint. Clearly, this community has different environmental priorities than those of Boston at large or the state.

The risk comparison carried out by the SAB in 1990 was not a one-time interpretation. Indeed, some of the information in Table 4.3 may now be outdated and the SAB has recently started a new study to rank environmental risks. It is possible, even likely, that some of their conclusions will differ from those of 1990.

Population and consumption

An issue crucial to all environmental concerns that the SAB only marginally addressed is population. In 1992, the United States National Academy of Sciences (NAS) and its British equivalent, the Royal Society of London, put out their first joint statement: "If current predictions of population growth prove accurate and patterns of human activity on the planet remain unchanged, science and technology may not be able to prevent either irreversible degradation of the environment or continued poverty for much of the world." One academy member commented, "Scientists in general are doing a lot more talking about global warming and ozone depletion than they are about the basic forces that are driving those things." More environmental organizations also recognize the seriousness of population growth. The greatest population increase is taking place in underdeveloped countries. However, the United States, unlike many Western European countries where population has stabilized, continues to have a higher birth than death rate and its population is also increasing through immigration. Many third world representatives resent the United States emphasis on population. They stress the relationship be-tween high levels of consumption and environmental degradation. On this ba-sis, it is the United States and other industrialized societies that are the major contributors to environmental degradation. In their book, *Healing the Planet,* Paul and Anne Ehrlich estimated that the average American has roughly 50 times the environmental impact of a Bangladeshi. A major challenge that Americans may confront is how we can cut consumption without disrupting our whole economic system. Another is how we can cut consumption in an equitable manner.

DISCUSSION QUESTIONS

1. Examine Table 4.3 again.
   (a) What risks in your state or community may be different than those shown in the table?
   (b) Can risk assessment be based only on scientific data? Explain your position.
   (c) Think ahead 20 years and assume that greatly improved sci-entific data are available. Will nonscientific factors still be im-portant? Provide examples to support your viewpoint.
2. (a) What environmental risks to human health and the environ-ment most concern you, and why do you consider them to be high-priority risks?
   (b) What risks to human health are of less concern to you and why?
3. Environmental risks also may adversely affect quality of life.
   (a) What does this term mean to you?
   (b) What environmental problems do you believe pose high risks to quality of life?

FURTHER READING

Chemical risk assessment

Hrudey, S. E., and Krewski, D. 1995. Is There a Safe Level of Exposure to a Carcinogen? *En-vironmental Science & Technology,* 29(8), 370A-75, Aug.

Glickman, T. S., and Gough, M. (eds.) 1990. *Readings in Risk.* Washington, D. C.: Resources for the Future.

Kolluru, R. V. 1991. Understand the Basics of Risk Assessment. *Chemical Engineering Progress,* 87(3), 61–67, Mar.

Patton, D. E. 1993. The ABCs of Risk Assessment. *EPA Journal,* 19(1), 10–15, Jan./Feb./Mar.

Upton, A. C. 1994. Science and Judgment in Risk Assessment. *Health & Environment Digest,* 8(7), 53–55, Nov.

Comparative risk assessment

Cleland-Hamnett, W. 1993. The Role of Comparative Risk Analysis. *EPA Journal*, 19(1), 18–23, Jan./Feb./Mar.

Cooper, W. 1992. Ranking Environmental Risks. *Health & Environment Digest*, 6(6), 1–3, Oct.

Ehrlich, P. R., and Ehrlich, A. H. 1991. *Healing the Planet*. Reading, Massachusetts: Addison-Wesley.

U.S. EPA, 1990. *Reducing Risk: Setting Priorities and Strategies for Environmental Protection: The Report of the Science Advisory Board to W. K. Reilly, Administrator*, SAB-EC-90–021, Sept.

5

AMBIENT AIR POLLUTION

Ambient air is outside air that is free to move; it is the open air around us. Ambient air pollution is also referred to as *tropospheric* or *ground level air pollution*. The troposphere, the lowest region of the earth's atmosphere, begins at the earth's surface. Three classes of ambient pollutants are covered in this chapter – criteria air pollutants, hazardous air pollutants (HAPs), and volatile organic chemicals (VOCs). These are the three categories specified by the CAA. See Table 5.1 for some characteristics of these pollutants and note the importance of fossil fuel combustion in their formation.

CRITERIA AIR POLLUTANTS

The term *criteria air pollutants* is from the 1970 Clean Air Act, which required EPA to set standards that would protect human health and welfare from hazardous air pollutants. To do this, EPA used specific criteria (characteristics and potential health and welfare effects of pollutants) to identify the most serious pollutants and to set standards for them. It identified six pollutants that account for the large majority of air pollution both in the United States and worldwide. These are carbon monoxide (CO), sulfur dioxide (SO_2), nitrogen oxides (NO_x), ozone (O_3), particulates (PM_{10}), and lead (Pb). Recall that the higher the dose of a particular chemical to which a person, animal, or plant is exposed, the greater is the possibility of an adverse effect. Now consider criteria air pollutants: They are often present in amounts high enough to exert adverse effects. Air pollution in Los Angeles and Mexico City often causes painful breathing, eye irritation, and headaches and also exerts adverse effects on the growth of trees and crops. Exhaust from motor vehicles accounts for about half of criteria pollutant emissions. Other anthropogenic sources are coal-fired electric power plants and industry. It is important to note that all these sources involve combustion, largely combustion of fossil fuels. Because each criteria pollutant is important, each is discussed individually.

80

TABLE 5.1 Ambient air pollutants

POLLUTANT	CHARACTERISTICS OR EXAMPLES
CRITERIA POLLUTANTS	SIX MAJOR POLLUTANTS FOR WHICH AMBIENT AIR STANDARDS HAVE BEEN SET TO PROTECT HUMAN HEALTH AND WELFARE.
Ozone (O_3)	A major component of photochemical smog formed from NO_x, VOCs, and oxygen in presence of sunlight and heat. Motor vehicles are a major generator of NO_x and VOCs.
Carbon monoxide (CO)	Produced by combustion of fossil fuel and biomass. By itself, CO represents more than 50% of all air pollutants. Motor vehicles are the major source of CO, especially in cities.
Sulfur dioxide [a] (SO_2)	An acid precursor converted to acid under moist conditions or to sulfate in dry. Both the acid and sulfate are particulates and major components of haze. Fossil fuel burning power plants produce 75%.
Nitrogen oxides [a] (NO_x)	Acid precursors converted to acid under moist conditions or to nitrate in dry. Both acid and nitrate are particulates and components of haze. In cities, motor vehicles generate about 60% of NO_x.
Lead [a] (Pb)	Emitted as particulates from metal mining and processing facilities, and by combustion processes.
Particulates (PM_{10})	Solid particles composed of one or several chemicals. There are many sources. Those from combustion processes pose the highest concern.
HAZARDOUS AIR POLLUTANTS (HAPs)	THE 189 HAPs DO NOT HAVE AMBIENT AIR STANDARDS. INSTEAD, EMISSIONS CONTROLS ARE USED. ABOUT 70% ARE ALSO VOCs.
Organic	Examples are benzene, formaldehyde, vinyl chloride.
Inorganic [b]	Examples are asbestos and metals (such as cadmium and mercury).
VOCs	ORGANIC CHEMICALS THAT EVAPORATE EASILY. SOME SIGNIFICANTLY CONTRIBUTE TO SMOG.

[a] A portion of this pollutant is converted to particulate form.
[b] A number of ambient air pollutants are emitted as particulates (example, metals).

Ozone

Ozone (O_3), with its three oxygen atoms, is related to molecular oxygen, O_2, which has two. Molecular oxygen is essential to life, but ozone is a more reactive chemical than oxygen and often adversely affects both plant and animal life. Many of us know the odor of ozone from lightning storms or have noticed it around equipment such as an improperly maintained photocopier. Ozone is a summer pollutant. It is found in smog, which is often called photochemi-

BOX 5.1

Beginning in this chapter, four questions are asked about the pollutants discussed:

1. What are the pollutants of concern?
2. Why are they of concern?
3. What are their sources and what is the route of human exposure? (See note below.)
4. What is being done to reduce sources of the pollutants?

Note: The source of a pollutant and the route of exposure to it are often different. Two examples demonstrate this: (1) PCBs: A major source of PCBs in the environment until the mid-1970s was industrial discharge into waterways, but the major route of human exposure to PCBs is fish consumption. This is explained as follows: PCBs released into water bodies were bound to sediments, which serve as a *sink* for these long-lived contaminants. Small aquatic organisms take up and concentrate PCBs from the sediment. These organisms with their contaminating PCBs are eaten by fish, and the PCBs concentrate in the fish fat. Birds and mammals, including humans, eat the contaminated fish and are thus exposed to PCBs. (2) Chlorinated dioxins: The major source of dioxins is incinerator emissions, but the major route of human exposure to dioxins is meat and dairy product consumption. This happens as follows: The dioxins that incinerators emit settle on vegetation that is eaten by cattle and other animals. The dioxins concentrate in the fat of the animal's meat and milk. Humans consume these products and are thus exposed to dioxins. A smaller source of dioxin is discharges into waterways. In this case, as with PCBs, fish are the route of human exposure.

cal smog because of the role the summer sun plays in its formation. Ground level ozone, tropospheric ozone, is the same chemical found in the stratosphere. But remember that in the troposphere, excess ozone is the problem, whereas in the stratosphere, ozone depletion is the problem.

Why is ozone of concern? Ozone is an oxidizing pollutant that is often present in the ground level environment at concentrations known to have deleterious health and ecological effects. Its acute effects are to irritate the eyes, nose, throat, and lungs and decrease the ability of the lungs to function optimally. In studies with young adults, ozone at 0.2 ppm caused inflammation of the bronchial tubes and of tissue deep within the lungs. Exercising people are particularly susceptible to ozone's adverse effects, as are those with asthma or bronchitis. Ozone may also increase susceptibility to infection. Chronic exposure to ozone may permanently damage the lungs. Hospital admissions for

TABLE 5.2 Ozone levels in ambient air

CONCENTRATION (PPM)	AIR QUALITY
0.00-0.06 ppm	Good
0.06-0.12 ppm	Moderate
0.12-0.20 ppm	Unhealthful [a]
0.20 plus ppm	Very unhealthful

[a] On days with unhealthful levels of ozone, children, the elderly, and those with repiratory problems such as asthma should reduce outdoor activities. On very unhealthful days, eliminate outdoor activities. Healthy individuals exercising outside should do so in early morning hours before ozone levels begin to climb. *Source:* Information from *Northeast Aireport*, Spring 1994, p. 7.

asthma are higher when ozone levels are high, although this effect is not solely due to ozone. The standard, that is to say, the enforceable limit, that EPA set for ozone is 0.12 ppm. This level was set to protect human health, but many argue that human health is adversely affected at levels below 0.12 ppm. However, more than a third of Americans live in areas out of compliance with even 0.12 ppm. Los Angeles has 30 to 40 days a year above 0.20 ppm. Table 5.2 notes the relationship between ozone levels and human health. Most ground level ozone is anthropogenic, but ozone is also found naturally in areas remote from human activity at levels of 0.02–0.05 ppm.

The ozone levels in Table 5.2 refer to adverse effects on humans, but plants and trees are even more sensitive. As early as the 1940s, increasing ozone levels around Los Angeles were observed to damage vegetable crops greatly. Ozone is primarily generated in urban areas, but having a life span of a day or more, it spreads over large regions of the countryside. Estimated crop losses from ozone pollution are 5% to 10%. Sensitive crops are damaged at 0.05 ppm, whereas more resistant ones can withstand 0.07 ppm or higher. Depending on the geographic locale, an estimated 35% of crops are grown in areas where ozone levels exceed 0.05–0.07 ppm in the United States and worldwide. If current trends continue, up to 75% of the world's crops will be grown in areas with damaging ozone levels by the year 2025. Ozone also adversely affects tree growth, and many foresters consider ozone to be the air pollutant most damaging to forests. In areas that are subjected not only to ozone, but to acid rain and other air pollutants as well, the combination of pollutants may be more damaging than the effects of any one alone.

What are the sources of ozone? Ozone is not typically emitted as ozone. Rather it is formed from the ozone precursors, VOCs, and nitrogen oxides

(NO_x). In the summer's heat and the sun's strong UV rays, VOCs and NO_x react with atmospheric oxygen to form ozone. Motor vehicles are a major source of ozone precursors. Because of the way that ozone is formed, you may correctly guess that the ozone level in a city builds up over the course of a summer day. Ozone level is typically low early in the morning. Then motor vehicle exhausts from morning traffic increase levels of NO_x and VOCs. At the same time, the day becomes warmer and the sun's UV rays stronger. The stage is set for ozone formation.

How is ground level ozone being reduced? For a discussion of specific efforts being made to reduce emissions of ozone precursors, see the sections on nitrogen oxides and VOCs. The results of these efforts, on which billions of dollars a year is spent, have been slow and spotty. A major reason is our inability to control motor vehicles, which, in urban areas, emit up to 60% of NO_x and about 40% of the VOCs. Vehicle emissions – in response to CAA regulations – have dramatically decreased in the past 20 years, 90% or more per gallon of fuel burned. Nonetheless, total emissions remain high. This is true because the number of motor vehicles has greatly increased, as has the number of miles each is driven. Poorly maintained vehicles, even when quite new, have much greater emissions than well-maintained ones. Another difficulty in controlling ozone formation is that reducing either VOCs or NO_x does not necessarily lead to an equal reduction in ozone because of the complex reactions involved in its formation. On the brighter side, the number of U.S. cities exceeding the federal ozone standard fell from 97 in 1990 to 56 in 1992. It was recently found that certain VOCs contribute much more to ozone formation than others. This knowledge may lead to efforts to reduce specific VOCs – formaldehyde is one major culprit – that most contribute to ozone formation. As the situation now stands, a reformulated gasoline may emit smaller amounts of VOCs but may actually contribute more to ozone formation than a gasoline that emits more VOCs that are less active in ozone formation.

Carbon monoxide

Carbon monoxide (CO) is a colorless, odorless, flammable gas, which is a product of incomplete combustion. If carbon were completely oxidized during burning, complete combustion to carbon dioxide would occur and carbon monoxide would not be a problem. It is important not to confuse carbon monoxide with carbon dioxide. Carbon monoxide (CO) is an incomplete combustion product and can be toxic even at low concentrations, whereas carbon dioxide (CO_2) is a complete oxidation product.

Why is carbon monoxide of concern? All by itself, carbon monoxide accounts for more than 50% of air pollution nationwide and worldwide – it is a pervasive pollutant. Worldwide, hundreds of millions of tons are emitted each year. Moreover, hundreds of Americans die each year from carbon monoxide poisoning in homes and other enclosed spaces (see Chapter 13). Many thousands more suffer from carbon monoxide-related illnesses, which include headaches, dizziness, and drowsiness. At higher carbon monoxide concentrations there is a wider range of symptoms. Recent reports indicate that even very low levels of outdoor carbon monoxide aggravate certain heart problems. Up to 11% of hospital admissions for congestive heart failure in elderly people may be caused by carbon monoxide. If you think about how carbon monoxide works, you may understand how this gas adversely affects the heart and other organs. In the normal situation, the iron atom in the blood protein, hemoglobin, picks up oxygen from the lung and transports it to the body's cells. There the hemoglobin releases oxygen and picks up the waste gas carbon dioxide, which it transports back to the lungs and releases. After releasing carbon dioxide, it picks up more oxygen. Carbon monoxide has 200 times greater affinity for the iron in hemoglobin than does oxygen and interrupts this cycle by displacing oxygen. The result is a lowered amount of oxygen reaching the heart, which can lead to heart failure in sensitive people. Carbon monoxide also has other adverse effects in the body; for example, it interferes with the oxygen-carrying proteins in muscles.

What are the sources of carbon monoxide? Carbon monoxide is formed whenever a carbon-containing material is burned. Thus, there is potential exposure to carbon monoxide in any place that combustion occurs. In American urban areas, 70% or more of carbon monoxide emissions are from motor vehicles. Vehicle drivers and traffic control personnel are sometimes highly exposed to carbon monoxide, as are mechanics working in garages and parking garage attendants. Cigarette smoke contains carbon monoxide; if individuals who are exposed to carbon monoxide on the job also smoke, the additional exposure enhances the chances of adverse effects. In addition to motor vehicles, sources of carbon monoxide include facilities burning coal, natural gas, or biomass (vegetation). Biomass combustion can be a significant source of exposure in rural areas or in underdeveloped countries where it is burned for cooking, heating, and even light. Atmospheric oxidation of methane gas and other hydrocarbons also produces carbon monoxide.

How are carbon monoxide emissions being reduced? About half of the motor vehicle carbon monoxide emissions in this country are produced by only 10% of the vehicles. Efforts are being made to find and remove these vehicles from the road. Car and truck owners need to maintain their vehicles so that they operate as cleanly as they were designed to operate. Other measures to control carbon monoxide emissions include requiring facilities that burn fossil fuels or wood to maintain high burning efficiencies and prohibiting open burning of trash and garbage. In the 1950s, atmospheric carbon monoxide levels began to increase. Then the 1970 CAA and its 1977 and 1990 amendments mandated increasingly stringent controls on motor vehicles. The result was that emissions decreased 24%, as compared to those of 1970. More recently, oxygen-containing fuel additives are now added to gasoline in some U.S. cities to enhance burning in winter, when engines tend to run less efficiently. What happened in 20 cities is illustrative of how oxygenated fuels can make a difference: In the winter of 1991–92, these 20 cities exceeded EPA's carbon monoxide standard of 8 ppm for 9 hours on 43 different days. However, in 1992–93, after oxygenated fuel was introduced, these cities exceeded the standard on only 2 days. Western Europe has also tightened carbon monoxide emission standards and levels have also fallen there.

Sulfur dioxide

Sulfur dioxide (SO_2) is a colorless gas with a sharp odor that accounts for about 18% of all air pollution, making it second only to carbon monoxide as the most common urban air pollutant.

Why is sulfur dioxide of concern? Sulfur dioxide gas reacts with moisture in the eyes, lungs, and other mucous membranes to form strongly irritating acid. Because it is so reactive, about 90% of sulfur dioxide is removed in the upper respiratory tract. Exposure to sulfur dioxide can aggravate already existing respiratory or heart disease. Sulfur dioxide and other sulfiting agents in foods can also trigger allergic reactions and asthma in sensitive individuals. As is the case with ozone, exposure to low concentrations of sulfur dioxide can damage plants and trees.

These direct effects of sulfur dioxide gas on humans and vegetation represent only one aspect of this tremendously important pollutant. If moisture is present in the atmosphere, sulfur dioxide is converted into sulfuric acid or, if conditions are dry, into sulfate particulates. The tiny – only 0.1 to 1 μm in diameter – sulfuric acid and sulfate particulates form *aerosols* (an aerosol is a gaseous suspension of fine solid or liquid particles). The aerosols contribute to the adverse health effects of smog and haze, and also play a serious role in haze (discussed in the section on particulates). Sulfuric acid and sulfate are likewise directly involved in three serious global change problems. Acidic deposition (acid rain) is one of these. The second is stratospheric ozone depletion, whereby sulfate particles in the stratosphere provide surfaces on which ozone-destroying reactions occur. A third major effect is the antiwarming influence they exert in global climate change. These will all be discussed in Chapter 6 (also see Table 6.1).

What are the sources of sulfur dioxide? Worldwide, more than two times as much sulfur dioxide arises from anthropogenic sources as from natural sources. However, in the highly industrialized Northern Hemisphere, including the United States, human activities produce five times more sulfur dioxide than natural sources. In 1985, electric utilities burning fossil fuels produced 75% of the anthropogenic sulfur dioxide in this country, about 23 million tons. Those utilities that burn high-sulfur coal produce the most. Another source, producing 15% to 20% of sulfur dioxide emissions, is metal smelters. This happens because many metals are present in ores as sulfur-containing compounds. Smaller sources are pulp and paper mills and oil refineries. Petroleum contains sulfur, but it can be more easily removed from petroleum than from coal, so motor vehicles account for only a small percentage of sulfur dioxide emissions. Although most sulfur dioxide is emitted by human activities, there are many natural sources, including marine plankton, sea water, bacteria, plants, and geothermal emissions. Volcanic eruptions are a major but periodic natural source. In 1991, about 20 million tons of sulfur dioxide was ejected into the atmosphere by the Filipino volcano Mt. Pinatubo. This huge quantity of sulfur dioxide was believed to be responsible for cooling the earth's climate for several years thereafter.

How are sulfur dioxide emissions being reduced? The 1970 CAA and its 1977 amendments mandated controls on sulfur dioxide emissions from coal-burning electric power plants and other sources. As a result, sulfur dioxide emissions in 1985 were 30% lower than in 1970. EPA reports that just between 1984 and 1993, urban sulfur dioxide levels fell 26%. In addition to end-of-pipe controls on sulfur dioxide emissions, the P^2 measure of using fuels containing lower amounts of sulfur also contributed to this reduction. However, reduced sulfur dioxide emissions have not prevented a troubling problem. As a result of sulfuric acid and sulfate particles visibility in eastern national parks is worse, not better. Mandates from the 1990 CAA Amendments will cut emissions further, to 10 million tons below the 1980 level. International efforts are also under way to cut sulfur dioxide emissions, especially in Europe.

Nitrogen oxides

There are several nitrogen oxide (NO$_x$) gases – nitric oxide, nitrogen dioxide, and nitrous oxide – collectively referred to as NO$_x$ (pronounced "knocks"). Each NO$_x$ gas is composed of varying combinations of nitrogen and oxygen.

Why are nitrogen oxides of concern? They account for about 6% of United States air pollution. Like sulfur dioxide, nitrogen oxides contribute to a number of problems. Direct exposure to nitrogen oxides irritates the eyes and lungs and can lower resistance to infection. A strong link has recently been made between exposure to nitrogen oxides and asthma. Like sulfur dioxide, nitrogen dioxide is poisonous to plant life. Also like sulfur dioxide, nitrogen oxides – to a lesser extent than for sulfur dioxide – are involved in the three global change problems to be discussed in Chapter 6. However, nitrogen oxides have important effects that sulfur dioxide does not. They are major contributors to the formation of ground level ozone, whereas sulfur dioxide is not. Also, the association of nitrogen oxides with plants is complicated by the fact that nitrogen is an especially important plant nutrient, and so there are sometimes beneficial effects.

What are the sources of nitrogen oxides? Nitrogen oxides are formed through the reaction of individually benign atmospheric nitrogen and atmospheric oxygen during high-temperature combustion. They are also formed from the burning of nitrogen compounds naturally present in fuels. As with carbon monoxide, the major anthropogenic source of nitrogen oxides is motor vehicles, which in urban areas contribute up to 60%. Burning gasoline at high temperatures is desirable as it increases burning efficiency and reduces carbon monoxide and hydrocarbon emissions. Unfortunately, high temperatures promote the reaction between atmospheric nitrogen and oxygen that increases nitrogen oxide levels. Another 25% to 30% of the anthropogenic nitrogen oxides in this country is emitted from electric utilities. Smaller amounts are emitted by other industrial combustion processes. Natural nitrogen oxide sources include lightning and volcanoes. Microorganisms decomposing vegetation also produce and release nitrogen oxides. One NO_x gas, nitrous oxide, is emitted by soil bacteria in especially large amounts after nitrogen fertilizer has been added to the soil.

How are nitrogen oxide emissions being reduced? Nitrogen oxides are harder to control than is sulfur dioxide. Because they are directly formed by a reaction between atmospheric oxygen and nitrogen at high temperatures, there is also sometimes a conflict between burning at high temperatures to achieve efficient combustion and the need to minimize nitrogen oxide formation. Although less effectively than was the case for sulfur dioxide, the mandates of the 1970 CAA and its 1977 Amendments reduced nitrogen oxide emissions. In 1985, emissions were down 8%–16% as compared to 1970. The 1990 CAA Amendments will cut nitrogen oxide emissions further to 2 million tons below the 1980 level.

Particulates

Whereas the gases nitrogen or oxygen are homogeneous parts of the air, particulate matter is suspended in air, sometimes as fine aerosols. Particulates – also referred to as *particulate matter* or sometimes just as *particles* – account for about 10% of U.S. air pollution. The term *particulates* can be confusing for several reasons. One is their composition. All the other criteria air pollutants are specific chemicals, carbon monoxide, ozone, sulfur dioxide, nitrogen oxides, and lead. However, particulates vary in chemical composition. One particulate may contain only one chemical such as sulfate or lead. Another may contain a number of components, sulfates, nitrates, metals, or dust or biological matter from many sources. The size of particulates also varies greatly. Further-more, they can be either dry or liquid. Very fine particulates of either liquid or solid that form a semistable suspension in air are referred to as aerosols; for example, sulfuric acid can form an aerosol. Complicating the issue of particulates further, gases are sometimes converted to particles.

Examine Table 5.1: Although particulates are listed as one of the six criteria air pollutants, this can be confusing because a number of other ambient air pollutants noted are also particulates or can be converted to particulates after emission. Some sulfur dioxide and nitrogen oxide emissions are converted in the atmosphere to particulates; sulfate particulates are an important portion of the tiny particulates of most concern. With the exception of some mercury, metal HAPs are emitted as particulates. The HAP asbestos is also a particulate pollutant. Also, some organic vapors, VOCs, can directly condense into partic-ulates.

Why are particulates of concern? In the past, before protective work place laws were passed, particulate matter took a terrible toll on American workers. High levels of silica dust caused silicosis in miners. Coal miners suffered black lung disease from breathing in coal dust. Textile workers suffered brown lung disease from inhaling cotton dust. Other workers, exposed to airborne as-bestos, had asbestosis, lung cancer, or mesothelioma. All these diseases are disabling or deadly. In many instances of occupational exposures, workers in-haled large amounts of particles of all sizes. Large-particle pollution has been much reduced both in ambient air and in the work place. The particulates of most concern today are very tiny. EPA regulates particles that have a diameter of 10 μm or less; these are referred to as PM_{10}. Sulfate particles and some soot and dust particles can be as small as 0.01 μm. Whereas larger particles of dirt, dust, or pollen can be caught in the nose, throat, or windpipe, from which they can be sneezed, coughed, swallowed, or spit out, the tiny PM_{10} particles can be inhaled so deeply into the lungs that they are much harder to expel. Particles that reach the lung alveoli, the tiny air sacs where oxygen is ex-changed with carbon dioxide, are a special problem. They may directly irritate the alveoli or else be absorbed into the bloodstream and exert adverse effects elsewhere in the body.

The idea of a particulate or particle seems very innocuous – a tiny piece or speck – not a concept to inspire alarm. Yet even low levels of particulates ad-versely affect breathing and can cause lung damage. Not one, but a series of epidemiologic studies showed a direct relationship between air particulate levels and respiratory problems. Particulate air pollution is associated with in-creases in emergency room visits and hospitalizations for respiratory prob-lems. Individuals at special risk from particulates include older people, small children, and those suffering from asthma, bronchitis, and emphysema. There is a reported relationship between particulate pollution and death rates among sensitive individuals. Even more significantly, a Harvard study found that, in cities with the most very fine particles in the air, a person's average life span was shortened 1 to 2 years. Even given the limitations of epidemiologic studies discussed in Chapter 3, these studies are impressive. However, some epidemiologists believe that the hazard is not due to particulates alone, but to air pollution more generally – particulates, carbon monoxide, plus ozone. Recall that ozone and carbon monoxide are also associated with increased hospital admissions under some conditions. You are probably not surprised that inhaled particulate matter can cause respiratory problems. Perhaps more surprising is a 1995 Detroit study that showed a correlation with heart problems: With each 100-μg/m^3 increase in air particulates, hospital admissions of people with the cardiac condition problem ischemia rose about 6% and for people with heart arrhythmias, about 8%.

In addition to health impacts, particulates contribute to the haze or smog seen in many cities, which also spreads far into rural areas. Haze has reduced visibility in America's western national parks as much as 50%. Eastern parks, which are exposed to a greater number of emission sources, are most affected –

BOX 5.2

The two meanings of smog

1. The most general definition of *smog* is visible air pollution. The word was first used in 1905 to describe the combination of smoke and fog that sometimes totally obscured visibility in London. This smog resulted from sulfur dioxide, soot, and tarry materials released by the uncontrolled burning of high-sulfur coal. A similar smog, often severe, also occurred in the coal-burning eastern United States before its sources were reduced. This smog primarily affected the lungs, and severe episodes caused thousands of premature deaths as well as illness. This type of smog is still a serious problem in Eastern Europe and other locations where there is uncontrolled burning of coal.

2. In contrast, in the United States today, the word *smog* most often refers to photochemical smog. This smog results from the reaction of VOCs and nitrogen oxides in summer sunlight to form ozone, a major component of smog. In addition to ozone, smog contains other photochemical oxidants, including peroxyacyl nitrates (PANs) and nitrogen dioxide. It also contains the particulates, which make it "air that you can see." Acute effects of photochemical smog are eye irritation and headaches. The lungs are also affected; indeed, breathing can be painful and chronic exposure can damage the lungs.

visibility is down 80% as compared to that in the 1940s. In the 1980s alone, Shenandoah and Great Smoky Mountains national parks suffered a 40% increase in sulfate particles. Instead of a blue sky, there is a pale white haze or a gray fog composed of dilute sulfuric acid. The increase in sulfuric acid and sulfate hazes in the 1980s is puzzling and disturbing because sulfur emissions overall have been decreasing. More research is clearly needed to understand what is happening.

What are the sources of particulates? The major source of the tiny particles of most concern, PM_{10}, is combustion of all types in motor vehicles, electric power plants, or other industrial operations. Electric power plants are a major source of sulfur dioxide, which can be converted to the sulfate particulates that contribute heavily to haze. Power plants and other incinerators also produce fly ash, composed of very fine particles that contain many metals as well as silicon. Silicon dioxide is the major component of sand and window glass, but it is less benign when found in fly ash's tiny particles. Fly ash particles also absorb the dioxins formed during combustion. When combustion is very efficient, almost all the organic material present is converted to carbon dioxide and water so little soot (a particulate) is formed. When combustion is less efficient, greater quantities of soot particles are generated. Wood-burning stoves, which often burn inefficiently, may lead to unhealthy particle levels in locales with large numbers of these appliances.

Combustion processes are a major source of the tiny particulates of most concern, but there are many other sources. Rural areas generate airborne particles from burning biomass and from windswept dirt, fertilizer, dried manure, or dried crop residues. Particulates in coastal areas contain high levels of chloride, which can corrode local buildings and statuary. Construction sites release large amounts of dust. Like a number of other pollutants, particulates may be found at higher levels inside the house than out (see Chapter 13). Indeed, a recent United States EPA study provided the surprise information that – like the Peanuts cartoon character Pig-Pen – each of us moves in a cloud of tiny particles. These particles, released from clothing, carpets, furniture, and other items, are stirred up as we move around a house or office. Cooking activities also release particulates. Large indoor amounts of particulates can result from a fire left smoldering in a fireplace or cigarettes left to smolder. Indoor dust particles can contain almost any contaminant that is capable of becoming airborne within the home – molds, bacteria, pesticides tracked in with dirt on shoes, or particles of peeling leaded paint. Industrial work places, although much better controlled than in earlier years, may also have particulate matter in the air.

How are particulate emissions being reduced? Emissions of particles smaller than 10 μm in diameter (PM_{10}) are controlled to meet an EPA standard of 150 micrograms per cubic meter (μg/m³) of air. However – although many cities

BOX 5.3

The ubiquitous PAHs. Polycyclic aromatic hydrocarbons (PAHs) are long-lived pollutants that cling to sediments, bioaccumulate in fat, and are difficult to degrade. These characteristics make PAHs similar to PCBs and other polychlorinated chemicals, but PAHs do not contain chlorine. Rather, they are polycyclic hydrocarbons formed by the incomplete combustion of fuels.

- PAHs pose several health concerns. Breathed into the lung as fine particulates, they can cause respiratory distress. A number are known human carcinogens; benzo[*a*]pyrene is the best known of these. Recall from Chapter 4 that an excess risk of no greater than one in a million is a typical goal for exposure to human carcinogens. However, PAHs are so ubiquitous that even rural soils away from major highways sometimes pose an excess risk close to one in a million and PAH levels in urban soils have a 100 to 1,000 times greater risk. There are factors that may make this statement less alarming. One is that a significant proportion of PAHs may not be bioavailable, that is, although they can be chemically detected, they are so firmly associated with soil that absorption into the animal body is limited.
- Every year, more than 4 million tons of PAH particulates is emitted into U.S. air. Natural PAH sources include forest and grass fires and volcanic eruptions. But it is coal and gasoline burning by humans that has increased environmental levels; wood burning is also a significant source in some locales. Airborne PAH particulates settle into water; concentrated in its sediments, they are protected from the sunlight that could destroy them. Airborne PAHs also settle onto soil, food crops, and other vegetation. The route of 90% of human exposure to PAHs is from food consumption, especially of leafy vegetables and unrefined grains. Significant quantities also form on charcoal-grilled foods and on browned meat unless precautions are taken. To a lesser extent, PAHs also form on browned toast and baked goods. In homes with cigarette smokers, tobacco is the primary route of PAH exposure. Smoking one or more packs a day can double, and in some cases quintuple, a smoker's exposure to PAHs.
- PAHs are particulates and the controls exerted on their emissions are similar to those already described. They are one more reason for concern about our dependence on fossil fuels.

barely meet this standard – studies have shown associations between very fine particulates and increased respiratory problems and premature death rates at levels only one-third of the standard. In the near future, particulates of diameters 2.5 μm and less may be regulated. More effort at P^2, rather than more regulation, is, of course, desirable; however, this would involve a commitment to lessen our dependence on fossil fuels, a commitment many are not yet willing to make.

Lead

Lead, a highly useful metal, has been mined for thousand of years. And it has been known for thousands of years (from observing miners) that lead is toxic to the nervous system. It was intensively mined in the ancient world, so much so that recent studies of ancient ice layers in Greenland have shown that lead – from particles that were air transported from Greek and Roman mining sites – was present at four times background levels 2,000 years ago. However, it is in the twentieth century that lead has become so widely mined and used as to make it ubiquitous in the environment. The level of lead in modern human skeletons and teeth is at least a hundred-fold greater than the level found in pre-industrial age skeletons. Its concentration in recently formed Antarctic ice is four times greater than in ice formed prior to the industrial age, and lead concentration in recent coral shells is 15 times greater than in those deposited a century ago. Although we may think of lead in a form like a pipe or battery, it is as an airborne particulate that lead is of most concern. This important pollutant will be covered in more detail in Chapter 9.

Criteria air pollutants in the year 2000

Criteria air pollutant levels in the United States declined in the 1980s as a result of the 1970 CAA and its 1977 amendments. Most dramatically, between 1984 and 1993, the levels of lead in the air fell 89%. The level of sulfur dioxide fell by 26% in the same period, and that of both carbon monoxide and nitrogen dioxide by 12%. Even the recalcitrant ozone level fell in 41 U.S. cities, enough to meet the federal standard, although 56 cities still exceed the standard. Smog, of which ozone is a major component, dropped 12%. In 1990, the CAA Amendments tightened motor vehicle emissions even more and required some cities to use reformulated gasoline. Nonetheless, criteria pollutants remain major problems as the year 2000 approaches. Some of the remaining problems follow.

- Sulfur dioxide emissions are definitely down, but visibility in national parks and other locales, as a result of sulfuric acid and sulfate particulates, is

BOX 5.4

Time for new standards? In late 1996, EPA proposed new standards for
both ozone and particulates. While maintaining current controls for particles 10 μm and less, particulates below 2.5 μm in diameter, $PM_{2.5}$, would
also be controlled. If this proposal were adopted, it would affect electric
power companies and other industries that emit the very tiny particles
not yet controlled. Diesel-fueled vehicles also emit particles of this size.
Industry is combating the new proposal as being exorbitantly expensive
while also stating the belief that EPA has not shown that it is particulates – rather than other pollutants occurring with particulates (carbon
monoxide, sulfur dioxide and ozone) – causing the observed adverse
health effects. Some observers advocate that the United States work in
cooperation with other countries to develop an international standard
that will not place American industry at a competitive disadvantage.

The new standard that EPA proposed for ozone is 0.08 parts per million measured over 8 hours instead of the current standard of 0.12 parts
per million measured over 1 hour. A stricter ozone standard would primarily affect – not industry – but American counties with heavy motor vehicle traffic; many of these are out of compliance with even the current
standard. Ozone also presents a conundrum. Recall from Chapter 4 that,
for chemicals that are not carcinogens, a no observed adverse effect level
(NOAEL) is determined and then a safety factor applied to determine a
standard that is safe for human exposure. However, ozone appears to
cause biological responses right down to background levels – a NOAEL
cannot be determined. At the same time, the difference between the adverse effects seen in children playing outside at the current ozone standard as compared to what would happen with a new stricter standard is
believed to be small. Thus, the U.S. EPA's Clean Air Scientific Advisory
Committee noted that setting a standard for ozone is more of a policy
call than a scientific judgment.

worse not better. Increased controls mandated by the 1990 CAA may ease
this problem.

- Carbon monoxide also continues to pose concerns with adverse health effects being demonstrated at ever lower levels, at least for sensitive populations. Almost 12 million people live in cities that regularly exceed the federal
  carbon monoxide standard, primarily because of heavy motor vehicle traffic.
- Ozone remains a particularly troublesome pollutant. Health effects can be
  seen at levels below the standard and many cities remain out of compliance
  with even this standard. The most daunting task is controlling ozone pre-

cursor emissions from our 185 motor vehicles that travel trillions of miles a year. Quite aside from human health concerns, ozone affects crops and forests, often at levels below those affecting human health.

• Particulates are equally troublesome with adverse health effects observed at levels well below the federal standard. Although power plants burning coal are a major source, there are many other sources. Furthermore, several chemicals can coexist in a given particle; even knowing all the variables involved in particulate problems is difficult. The very fine particles that pose the greatest concern are the ones that are the hardest and most expensive to capture.

Criteria air pollutants remain a serious problem in the United States and other countries. Continuing population growth and development activities on a scale greater proportionately than population growth directly impair air quality. Human beings have adversely impacted local environments for thousands of years, but the number of people was small compared to the space available and people moved on to other locations. Today's large and growing population means we can no longer just move on.

DISCUSSION QUESTIONS

1. What criteria air pollutant(s) would most concern you in each of the following instances? Explain your choices.
   (a) You are a garage attendant or a traffic officer in a large city.
   (b) Air pollution from a nearby urban center reaches your farm.
   (c) A truck is parked near an air intake of the motel where you are spending the night. The truck contains perishables and its motor has been left idling.
   (d) You are a park ranger in the Great Smoky Mountains National Park.

2. Consider the proposed new standards for particulates and ozone. Remember that the tiny particulates of most concern are emitted by industry and diesel burning trucks. Conversely, the emissions resulting in most ozone formation comes from cars and light trucks. Many cities are out of compliance with even the current ozone standard. Are we willing as individuals to hold ourselves to standards as strict as those to which we hold industry?

HAZARDOUS AIR POLLUTANTS

Before discussing hazardous air pollutants (HAPs), it is important to emphasize once again that criteria air pollutants are also hazardous and often present at environmental levels where their potential hazards become real. For the student, a saving grace of the criteria pollutants is that there are only six. In contrast, many hundreds of chemicals fall into the more general HAPs category, also referred to as toxic air pollutants.

What are the pollutants of concern? The 1990 CAA amendments regulate 189 HAPs that Congress has identified as of special concern (see Table 5.3). Examples of inorganic chemical HAPs are asbestos and arsenic and the metals cadmium, mercury, and beryllium. Examples of organic chemical HAPs are benzene, chloroform, and formaldehyde. Radionuclides are also regulated. One HAP, coke oven emissions, is not an individual chemical, but a group of

TABLE 5.3 Examples of the 189 federally regulated hazardous air pollutants

ORGANIC CHEMICALS	REPRESENTATIVE SOURCES OR USES
Benzene	Gasoline, cigarette smoke
Toluene	Gasoline, motor vehicle exhaust, cigarette smoke, paints
Ethylene glycol	Automobile antifreeze, brake fluid
Methanol	Windshield antifreeze, solvent
Chloroform	Formed during water chlorination and other chlorine uses
Methyl bromide	Fumigant
Formaldehyde	Particleboard and plywood, building insulation, cosmetics
Parathion	Insecticide
Styrene	Manufacture of plastics, rubbers, adhesives, cushions
Vinyl chloride	Manufacture of plastics, new automobile interiors
INORANIC CHEMICALS	REPRESENTATIVE SOURCES OR USES
Asbestos	Fibrous mineral once widely used to fireproof materials
Arsenic	Nonmetallic element used in metal alloys and glass making
Metals	
Cadmium	Electroplating, Nicad batteries, pigment, plastic stabilizer
Chromium	Electroplating auto parts and bathroom fixtures, also chemical catalyst
Mercury	In thermometers, lamps, dental amalgams
Nickel	Electroplating, in alloys, chemical catalyst

chemicals driven off coal during its conversion to the coke used in iron and steel production. About 70% of the 189 HAPs are also VOCs, another category of pollutants discussed in this chapter.

Why are HAPs of concern and what are their sources? Potential adverse effects depend upon the toxicity of the particular HAP under consideration and the amount released. Because it is not feasible to consider all 189 chemicals, only examples are given here.

- The HAP benzene is one of the highest-volume chemicals produced in the United States and is also found in gasoline. It can irritate the skin and eyes, cause headache and dizziness, and, at levels formerly found in some work places, has been associated with leukemia and aplastic anemia.
- Another high-production HAP is formaldehyde, which is released from factories manufacturing furniture or pressed wood products. It can irritate the eyes and lungs and, at high doses, is an animal carcinogen. Some people develop severe allergies to formaldehyde.
- The HAP chloroform was used as an anesthetic for 100 years before its ability to damage the liver was fully appreciated. Exposure to chloroform can also damage the kidneys, and, at high concentrations, it is an animal carcinogen. Chloroform is released from sewage treatment plants and from pulp-bleaching facilities that use a chlorine-containing chemical.
- Cadmium, a metal HAP, is emitted by metal-refining operations and incinerators. Cadmium is a highly toxic metal that concentrates in plants, shellfish, and animal kidneys and liver.
- Mercury, another metal HAP, is a volatile liquid metal. It is especially toxic after conversion to methylmercury by bacteria. Methylmercury concentrates in animal tissues.

HAPs are often emitted locally at levels that can be of concern. Some are important only in the locale where they are produced. For example, chloroform emissions are important only near local facilities such as municipal wastewater treatment or pulp-bleaching facilities. However, some HAPs are widespread; for example, benzene is found anywhere gasoline is used to drive a motor vehicle or fill its gas tank or cut the lawn with a gasoline-powered mower. Benzene is also emitted in cigarette smoke. Each benzene source may only be local, but sources are ubiquitous.

Metallic HAPs present a special problem. Via wind currents, they spread far from initial sources; there are huge numbers of individual sources that emit metals – every motor vehicle that burns gasoline, every fossil fuel-burning power plant, and many industrial facilities. All the sources of these non-biodegradable pollutants added together constitute a problem.

How are HAP emissions being reduced? The 1990 CAA Amendments mandated the use of maximum available control technology (MACT) to reduce

HAP emissions. A voluntary program to reduce them had already begun before the passage of the 1990 CAA amendments. In the late 1980s, EPA pinpointed 17 HAPs as posing special concerns because they either were especially hazardous or were produced in large amounts. EPA asked facilities emitting large quantities of these 17 voluntarily to reduce emissions, by 33% by the end of 1992 and by 50% by the end of 1995. Over a thousand large companies reduced emissions accordingly. Making the public aware of a facility's emissions may also be a factor that has decreased emissions. The 1986 EPCRA law required many facilities to report how much of 320 chemicals (many of them HAPs) were released into air, water, or land. Many facilities, concerned about community reactions, worked harder to reduce these emissions than they might otherwise have. It is worthwhile to remember that the greatest source of human exposure to a number of HAPs – benzene, formaldehyde, and others – is within our own homes (see Chapter 13).

VOLATILE ORGANIC CHEMICALS

What are the pollutants of concern and why are they of concern? Volatile organic chemicals (VOCs) is a category that contains a very large number of easily evaporated organic compounds. About 70% of the HAPs discussed previously are among the VOCs and there are hundreds of others. To varying degrees, individual VOCs contribute to the formation of ground level photochemical ozone. VOCs also contribute to other problems; for example, many drivers and pedestrians experience headaches and other symptoms in reaction to the volatile hydrocarbons in motor vehicle exhausts. Sensitive individuals may react to exhaust fumes with attacks of asthma or other respiratory problems. Part of the illness may be directly due to VOCs and part to ozone-containing smog.

What are the sources of VOCs? An estimated 22 million tons of VOCs were emitted in the United States in 1985. There are a few large sources. A petroleum refinery or a chemical plant can be a very important local source. Motor vehicle exhausts are a major and pervasive source, emitting up to half of all American VOCs; internal combustion engines burn hydrocarbons inefficiently, and many products of this incomplete combustion are emitted in the exhaust as hydrocarbon VOCs. Hydrocarbons also evaporate when the gas tank is filled and when the vehicle is running, idling, or cooling.

There are many small VOC sources, including dry cleaners, auto maintenance shops, wood-drying or wood-painting operations, and even freshly painted houses. Restaurants and bakeries emit pleasant-smelling VOCs, but large bakeries also emit large amounts of ethanol formed by the action of yeast, which contribute to smog formation. Sewage treatment plants and composting operations often emit VOCs with objectionable odors. Humans are exposed to VOCs in busy city traffic and near the various facilities that emit them. In homes and work places, people are exposed to VOCs from paints, solvents, charcoal broiler starters, aerosol sprays, even deodorants and cosmetics. These home sources will be further discussed in Chapter 13.

Natural sources also emit large amounts of VOCs. Plants and trees emit hydrocarbon VOCs in hot weather. Even in a large city like Atlanta, trees are a significant source, and, in a heavily forested state like Maine, produce more than 90% of the VOCs. Among the many volatile compounds that trees emit are terpenes, responsible for the smell of pines; unfortunately, these pleasant VOCs contribute to the ozone formation that harms trees. However, trees do not emit the nitrogen oxides that interact with VOCs to form ozone – those are largely formed from human activities. Natural VOC emissions are not a problem by themselves but must be figured into any strategy to reduce urban ozone formation.

How are VOC emissions being reduced? In 1985, as a consequence of regulations resulting from the 1970 CAA and 1977 amendments, VOC emissions were about 30% lower than in 1970. The 1990 CAA amendments mandated reductions in HAP emissions, about 70% of which are also VOCs, as well as further reductions in motor vehicle VOC emissions. Regions with the worst air

---

BOX 5.5

Lawnmowers, tractors, and boats. The engines of the 185 million cars and trucks in this country produce up to a half of VOC emissions, half of nitrogen oxides, and 70% of carbon monoxide emissions. Americans also own or operate many other engines, including about 89 million lawn mowers and other pieces of lawn equipment plus "non-road engines" such as off-road vehicles, agricultural and construction equipment, and recreational marine engines. In total, these additional engines emit about 10% of the VOCs and 15% of nitrogen oxides. A walk-behind gasoline-powered lawn mower can emit as much air pollution in 1 hour as a new car does in 11 hours. This equipment has been unregulated, but EPA is beginning to require exhaust emissions standards for lawn mowers and for the diesel engines found in farm and construction equipment. Marine engines (outboard, inboard, and personal watercraft engines such as jet skis) are another significant pollution source. Not only do recreational vessel engines emit air pollutants, but about 30% of the fuel they consume is directly lost to water as unburned gasoline. Regulation of these sources will begin in 1998.

pollution problems had to commit to a 15% reduction in urban smog by 1996. To do this, certain locales, such as Southern California, took steps that include switching to less polluting motor vehicle fuels. California also has a law mandating that vehicle manufacturers begin to sell a certain number of less polluting vehicles and zero emission vehicles starting in 1998. In 1991, Congress passed the Intermodal Surface Transportation Efficiency Act, or ISTEA (pronounced "ice tea"). ISTEA provides a number of alternatives that states can choose among to reduce motor vehicle emissions. Options include highway lanes that can be used only by vehicles carrying more than one passenger and tolls that can be imposed on peak traffic hour drivers. Another option is asking employers to pay a bonus to employees who do not drive to work. See Chapter 12 for more information on reducing motor vehicle emissions.

DISCUSSION QUESTIONS

1. (a) Which air pollutants in your community or state cause you concern and why?
   (b) What are local sources of these pollutants?
   (c) Which originate partially or mostly beyond your state's borders?

2. Gasoline combustion in our 185 million motor vehicles is the single largest source of air pollution and consumes half of the country's petroleum.
   (a) Is knowledge of this information enough to contribute to changes in the way you buy, drive, and maintain your vehicle? Why?
   (b) If knowledge alone is not enough, what would have to happen to change your behavior?
   (c) Under what circumstances do you now walk, take public transportation, or ride a bike?
   (d) Under what circumstances would you be willing to use these options more often?

3. Some social critics have observed that the United States regulates pollutants end-of-pipe and encourages P^2 but continues to be blind to the root causes of environmental stress and damage: population and consumption.
   (a) Do you agree that environmental regulations will not be enough to limit pollution and environmental damage if population and consumption continue to increase? Support the position that population and consumption need not be reduced; then, support the position that they must be reduced.
   (b) Can consumption be lowered in such a way that good living standards (as you define them) can be largely maintained? If so, how?

4. Five ambient pollutants, carbon monoxide, sulfur dioxide, nitrogen oxides, particulates, and VOCs, account for about 98% of all air pollution in the United States. Fossil fuel combustion (largely from fueling motor vehicles and electric power plants) is a major source of all five.

    (a) Does this knowledge affect your opinion about our dependence on fossil fuels? Explain.

    (b) Should society be taking steps to reduce this dependence? If so, what steps?

    (c) What steps can you as an individual take to help reduce dependence on fossil fuels?

FURTHER READING

The Clean Air Act Amendments of 1990. 1991. *Health & Environment Digest*, 5(2), 1–5, Mar.

Fine Particles in Air Shorten Lives. 1994. *Environmental Health Perspectives*, 102(3), 274–75, Mar.

Holler, J. S. 1988. Volatile Organic Compounds. *Health & Environment Digest*, 2(5), 1–3, June.

Raber, L. R. 1997. Clean Air: Dollars Versus Lives. *Chemical & Engineering News*, 75(5), 28–30, Feb. 3.

6

AIR POLLUTION AND GLOBAL CHANGE

In a July 1996 issue of the *Boston Globe*, the columnist David Nyhan aptly described the linkage of air pollution to global change: "Wind, rain and radioactivity do not stop at the border for passport control, but go where they will. Pollution? Coming soon to a place near you. We're all Downwinders now." Air pollutants are blown across state and national borders, sometimes in significant amounts. Even when concentrations reaching distant points are low, environmentally persistent pollutants build up over time if emissions continue. For example, metal pollutants generated by European and Asian industry reach the Arctic and create a winter haze reported to cover an area equivalent to 9% of the earth's surface. Another important example is polychlorinated chemicals transported to the Arctic, which, because of its extreme cold, is a sink for these chemicals – they are unable to evaporate again. This chapter covers three major pollution issues that have regional or worldwide implications: acidic deposition (acid rain), stratospheric ozone depletion, and global climate change. Remember as you read that in addition to pollution other factors are also impacting the global environment. Important among these are continuing destruction of wildlife habitat and loss of biodiversity intertwined with increasing population and development. However, the United States environment is in good condition compared to that of many third world countries, where environmental degradation continues to worsen.

ACIDIC DEPOSITION

For many years, electric power plants, burning sulfur-containing fossil fuels, and smelters, processing sulfur-rich metal ores, produced local pollution that damaged local vegetation and water. To correct this problem, the 1970 CAA required power plants and smelters to build emission stacks 1,000 feet above ground level. The purpose of high stacks was to protect local communities. At the same time, it was expected that pollutants released at this height would be

---

BOX 6.1

Atmospheric deposition. Keep in mind that the term *acidic deposition* refers to only one type of pollutant deposited from the atmosphere; the term *atmospheric deposition* better describes actual conditions: not only acids, but many other substances, including metals and organic pollutants, are deposited onto forests and other vegetation, land, and water. Consider DDT and PCBs. There is no record that either was ever used in New Hampshire forests, but both are found in soils there, presumably airborne from other locations. Supporting this interpretation is the fact that greater amounts of these chemicals are found at higher elevations than at lower, especially on slopes facing prevailing winds. The amounts of DDT and PCBs deposited do not appear to threaten forest health. However, their very presence is one of many examples of pollutants airborne from sources within and beyond the United States. As discussed in this chapter, acidic pollutants can reach distant points in damaging amounts. Mercury and the pesticide toxaphene are other examples of air-transported pollutants that can be deposited at levels of concern.

---

so diluted in the atmosphere that they would cause no harm anywhere. Instead, a different problem emerged.

What are the pollutants of concern?

Sulfur dioxide (SO_2) and nitrogen oxides (NO_x) are precursors of acid rain. When moisture is present, chemical reactions in the atmosphere convert sulfur dioxide to sulfuric acid, and nitrogen oxides to nitric acid. These acids reach earth in rain, snow, and fog: wet acidic deposition. Sometimes acidic fog has direct contact with trees that grow at high altitudes or in coastal regions. If atmospheric conditions are dry, the precursors are converted to sulfate particles and nitrate particles, which settle by gravity as dry acidic deposition. Dry deposition is most likely to occur near emission sources. In the following discussion, sulfur dioxide emissions are emphasized, but keep in mind that nitrogen oxide emissions are also importantly involved. To a lesser extent, other chemicals contribute to acidic deposition. Carbonic acid is formed in the atmosphere from carbon dioxide. Organic acids such as formic acid and acetic acid are also present in the atmosphere as the result of natural processes and industrial activity. Certain other VOCs are also converted in the atmosphere to acidic forms. Sulfur dioxide not only plays an important role in acidic deposition, but is involved in all three global change issues discussed in this chapter (see Table 6.1).

TABLE 6.1   Sulfur dioxide and global change

ISSUE	ROLE OF SULFUR DIOXIDE
Acid rain	Sulfur dioxide emissions are converted in the atmosphere to sulfate particles or to sulfuric acid. These are the major contributors to acidic deposition.
Stratospheric ozone depletion	Sulfur dioxide from volcanic eruptions is injected into the stratosphere and converted to sulfate particles. Analogous to ice particles, sulfate particles provide surfaces on which ozone depleting reactions occur.
Global warming	Sulfur dioxide is converted to sulfate or sulfuric acid particles. These act as "antigreenhouse" substances by absorbing a portion of the sun's radiation, preventing it from reaching and warming the earth's surface.

Note: To varying extents, nitrogen oxide also has a role in each of these phenomena.

Why is acid rain of concern?

Acidic deposition, commonly called acid rain, was first described in 1852. Later, in the 1920s and 1930s, severe damage to trees and vegetation was described near metal smelters that released the gases sulfur dioxide and the acidic hydrogen fluoride. After high stacks were built in the 1970s, there was much less local damage; emissions from thousand-foot-high stacks were in-

---

BOX 6.2

pH. The term *pH* refers to how acid or how alkaline (basic) a solution is. A neutral solution has a pH of 7. As pH increases above 7, the solution becomes increasingly alkaline; as pH decreases below 7, it becomes increasingly acidic. Because the pH scale is logarithmic, each pH unit represents a 10-fold change in acidity or alkalinity. For example, a pH of 5 is 10 times more acid than a pH of 6 and a pH of 4 is 100 times more acidic than a pH of 6. Because of naturally occurring atmospheric acids, the pH of rain can naturally be low. However, a pH below 5.6 indicates that human activities are exerting an influence. Since 1965 the average annual pH of rain and snow in the northeastern United States has been pH 4.05–4.3, well below the cutoff of 5.6. Most of the United States east of the Mississippi River, including the Southeast, experiences acid rain to varying degrees. In a few cases, rain has been measured with a pH as low as that of vinegar, about 3, and lemon juice, about 2 (see Figure 6.1).

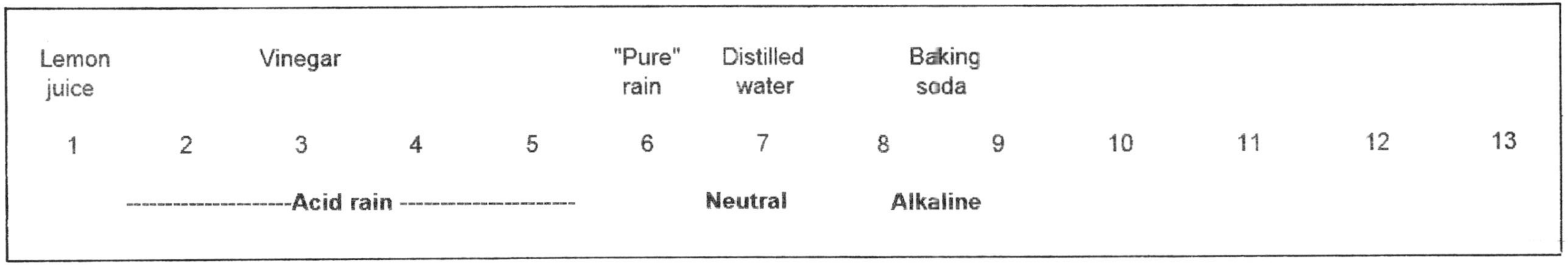

Lemon juice		Vinegar			"Pure" rain	Distilled water	Baking soda					
1	2	3	4	5	6	7	8	9	10	11	12	13

————————————Acid rain————————

Neutral   Alkaline

FIGURE 6.1. How "acid" is acid rain? Source: Adapted from *Meeting the Environmental Challenge: EPA's Review of Progress and New Directions in Environmental Protection,* December 1990.

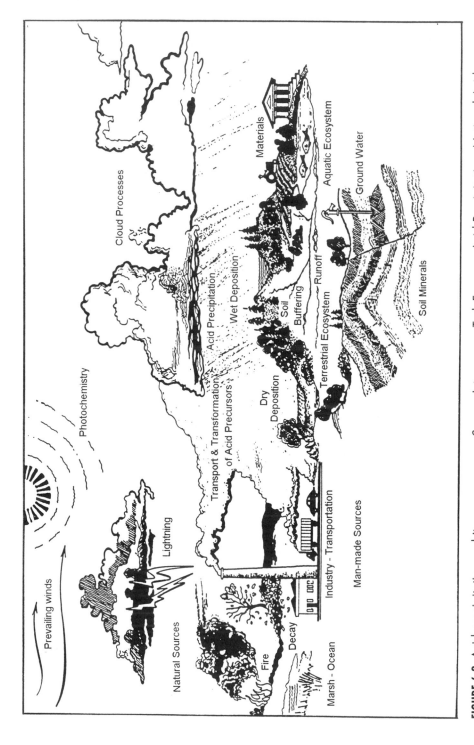

FIGURE 6.2. Acid precipitation and its consequences. *Source: Interagency Task Force on Acid Precipitation, National Acid Precipitation Plan, June 1982, p. 21.*

stead caught by the wind and carried away from the point of origin. But the acidic pollutants eventually settle out onto land and water. As compared to the level at the point of emission, they are very dilute. Alkaline soils can neutralize deposited acids over long periods. However, other soils may be naturally acidic or only weakly alkaline. These cannot continue to neutralize acid, and the results are acidified soil and acid runoff into nearby streams and lakes. Sometimes large amounts of acid trapped in winter snow run off into water bodies during spring snow melts. Acidic deposition also falls directly into water bodies. Between acid runoff and atmospheric deposition, water may become acidic enough to harm or kill fish and other aquatic organisms. Acid rain also leaches metals from soils more readily than rain with a higher pH. Leached aluminum is of special concern; aluminum runs off with rain into water bodies, where, especially if the water is already acidic, it stresses or kills fish and other aquatic organisms. In forests, too much soluble aluminum in the soil may interfere with uptake of the nutrients calcium and magnesium into tree roots. Trees stressed in this way may become more susceptible to disease or insect attacks. See Figure 6.2 for a representation of the sources of acid rain precursors and the systems that acid rain can impact.

To assess these concerns and to determine how much damage had already occurred, the United States government spent a half-billion dollars between 1980 and 1990 through the National Acid Precipitation Assessment Program (NAPAP). This ambitious study was carried out by teams of scientists throughout the country with the goal of gaining an understanding of acid rain and its effects. Many problems complicated the study: One is that rain does not have a "normal" pH. Acidic and alkaline substances are naturally present in air and affect the pH of rain to degrees that vary with local conditions. Rain in the eastern United States is often naturally acidic. The pH of water bodies also varies, as some are naturally acidic. So scientists studying water bodies needed to know their historical pH. Similarly, an understanding of whether acid rain was adversely affecting forests was complicated by the fact that many other stresses also affect trees. Among these are drought, temperature extremes in winter and summer, insufficient nutrients in the soil, insect attacks, fungal infections, and ground level ozone. Researchers had to struggle with the question of whether acidic deposition was adversely affecting forests above and beyond other stresses. In 1990, NAPAP investigators presented a report that concluded that acidic deposition demonstrated the adverse effects discussed later.

Effects on water and aquatic life. Acidic deposition had caused some surface waters to acidify. Fish populations and other aquatic life, including snails and crustaceans, were adversely affected in about 10% of eastern lakes and streams. Water bodies in the Adirondacks, New England, and the Middle-Atlantic states were especially vulnerable. Some observed adverse effects on

BOX 6.3

Forest dieback and decline happen every 50–200 years, brought about by a combination of stresses and old age. Air pollution is an added stress. Very low levels of air pollution may have no adverse effects. Intermediate levels may harm specific trees or species, but not in an obvious way. High levels lead to obvious disease or death of some trees. American foresters and air pollution experts believe that ozone is the air pollutant that most severely stresses trees. Heavy metal pollution is considered second in importance and, as of the early 1990s, acidic deposition was rated third. Continued efforts to reduce air pollution are important. Professor William Smith of Yale's School of Forestry and Environmental Studies observed, "The integrity, productivity, and value of forest and other wild systems are intimately linked to air quality. We must elevate considerations of environmental health to the same level as concerns for human health." Recall that EPA's SAB made the same recommendation in their report, that is, that ecological health should be treated as seriously as human health.

aquatic life were directly due to increased water acidity; others were indirect, as when aluminum, washed out of soil by acid rain, ran off into local water bodies. Aluminum is especially toxic to fish.

Effects on forests and soils. NAPAP investigators noted that red spruce trees growing at high elevations were in direct contact with acid clouds, which adversely affected them enough that they showed a reduced tolerance to winter cold. Beyond this, NAPAP scientists did not see evidence of widespread forest damage. They also concluded that some tree kills, previously blamed on acid rain, were caused by disease. Others objected to this conclusion and noted that, even if the immediate cause of tree deaths was disease, acidic deposition may have made the trees more susceptible to disease. Diseased trees may take many years to die, and the duration of the NAPAP study may not have been long enough to observe this. More generally, studies of pollutant effects on forests are complicated by the long lives of trees: Like chronic diseases that develop in human beings, adverse effects may show up only after a period of many years. Since the completion of the 10-year NAPAP investigation, studies of acid rain's effects on forests and soils have continued. One study, reminiscent of the NAPAP study of acid clouds on forests, showed harmful effects of acid fog on the needles of red spruce trees growing along the Maine coast. The acid degrades the wax on spruce needles. Damaged needles turn brown during the winter and fall from the trees. New needles grow in the spring, but if the damaging exposure continues, tree growth suffers.

Recently, more disturbing reports have emerged. A study described in a 1996 *Science* article indicated possible serious acid rain damage to the soil and trees of a New Hampshire experimental forest. Data had been collected on this forest for 30 years, but not all had been analyzed. Recent analysis of data on calcium revealed that the level of organically bound calcium in the soil had dropped to about half the value of 20 years earlier. Meanwhile, starting in 1987, this forest had stopped growing and suspicion centered on calcium depletion resulting from acid rain as a cause. Although calcium is replenished by rock erosion during weathering, replenishment could take decades even if no further calcium were lost. However, calcium depletion is continuing. To test the hypothesis that it is the depletion of calcium that stopped forest growth, calcium will be added to part of the forest to see whether growth resumes. The site to which calcium will be added is dominated by sugar maple, a species particularly sensitive to acid rain. Meanwhile, because calcium is an alkaline ion, less calcium in the soil means less ability to neutralize acid rain. In turn, this allows acid deposited on the soil to continue leaching metals from the soil. The authors of this study stressed that this problem has no quick fix. They also pointed out that, although the 1990 CAA amendments mandated that sulfur dioxide emissions be cut to 50% of 1980 levels by the year 2000, this reduction may not be enough.

Damage from acid rain to Central and Eastern European forests has been more severe than that in the United States; water and soil acidification is up to three times greater. Results from a German study indicated that the nitrogen in acid rain in one locale had come to pose a greater problem than sulfur. Nitrogen is an especially important nutrient that is often the limiting factor in tree growth – trees ordinarily absorb and use all they receive. However, over the years, acid rain resulted in so much nitrogen deposition that trees could not use it all. Many believe that excess nitrogen in German forest soil may be responsible for forest deaths that have occurred there. This is explained in the following way: Trees responded to the nitrogen in acid rain by growing abnormally fast. Already weakened by ozone and other pollutants, the trees were further weakened by rapid growth and eventually became unable to handle ordinary stresses like weather extremes or insect attacks. Worse, a situation similar to that described for the New Hampshire forest occurs; that is, the excess nitric acid leaches the alkaline calcium and magnesium from the soil. Rainwater then carries off the calcium and magnesium, depleting the soil of these nutrients. A vicious circle comes into play. As the alkaline calcium and magnesium are lost, soils become even less able to neutralize acid rain and become increasingly acidic, further damaging tree roots. Results of this study are also relevant to the United States. Until recently, sulfur in acid rain was the major concern, but as sulfur dioxide emissions are controlled, the less well controlled nitrogen oxides become an increasing concern.

Other effects of acid rain

- The corrosive effect of acid rain has resulted in increased rates of erosion of stone and metal structures in some regions. Stained glass windows and other cultural resources have also been damaged.
- The acid particulates sulfate and nitrate led to reduced visibility in the eastern United States and some western areas.
- Near major sources of acid precursor localized adverse effects on crop growth have been seen, but not at more distant points.
- In some cases, acid rain has beneficially affected plant growth because nitrogen and sulfur are plant nutrients.

What are the sources of sulfur dioxide and nitrogen oxides?

Coal-burning power plants in the midwestern United States pose a major problem. Prevailing winds carry the acid precursors emitted by these plants in a northeasterly direction to New York and New England. Acidic pollutants produced in the United States are also transported into Canada. To a lesser extent, acidic pollutants from Canada are also airborne into this country.

How are emissions of acid precursors being reduced?

The pH of rain in the northeastern states still averages 4.05 to 4.3, unacceptably low, but the reductions in acid precursor emissions mandated by CAA regulations are beginning to be noticed. Compared to the 1980 level, sulfur dioxide emissions from coal-fired utilities are down by more than 50% in both the United States and Canada. However, controls on nitrogen oxide emissions are just now being implemented. The U.S. Geological Survey reported that rain in 1993 was less acidic than in 1980 at 9 of 33 monitoring sites. Sulfate levels were down at 26 of these sites, but that of nitrate had decreased at only 3. Calculations indicate that controls already imposed by earlier versions of the CAA could have restored fish in the Adirondack lakes by the year 2030. The 1990 CAA amendments mandated an additional 10-million-tons-per-year reduction in sulfur dioxide emissions, which could restore fish by 2010. Few additional lakes are expected to acidify. Unfortunately, some effects of acid precipitation may not be reversible. Some soils in Middle-Atlantic states have reached their acid limit and cannot neutralize further acid; for them, cutting sulfur dioxide emissions in half will maintain the status quo but not reverse damage already done. Another troubling concern is the long-term effects of continuing nitrogen deposition on forest soil and tree growth. Yet another is increasing sulfate particle haze, which is occurring despite reductions in sulfur dioxide emissions.

Acid rain and international problems

More than 20 European governments have negotiated an agreement to lower sulfur dioxide emissions by as much as 87%. One means that will be used to accomplish this will be a switch from coal-fired to gas-fired electricity generation. Among the fossil fuels, coal typically produces the most sulfur dioxide and natural gas the least. Sulfur dioxide emissions in Western Europe have already been reduced 30% compared to those of 1980, but acid rain showed no corresponding decrease and there have been continuing serious effects on fish, forests, and monuments. A striking finding reported in 1995 may partially explain the lack of improvement: It was previously believed that acidic pollutants produced in the United States were washed out by rain before reaching Europe. Then a National Aeronautic and Space Administration (NASA) satellite, which carried a technology that can image pollution patterns from space by using laser light, found a large sulfate plume stretched from the eastern United States to Europe. This sobering finding has implications for the United States as well as Europe. Elsewhere in the world, acidic deposition problems are sometimes worse. Central and Eastern European countries cannot afford the technology to remove sulfur dioxide from the high-sulfur coal they burn. Their emissions of acid precursors not only seriously affect their own environment, but, carried in the wind to Scandinavian countries, cause acid rain problems there. In their own self-interest, Scandinavians are beginning to assist Eastern Europe with the technology to curb emissions. Similarly China produces sulfur dioxide emissions that cause problems in Japan. Much of the sulfur dioxide is from coal burned in many millions of Chinese homes for cooking and heating. The Japanese have developed a coal briquette, suitable for use in homes, which contains limestone that can partially absorb sulfur dioxide emissions.

DISCUSSION QUESTIONS

Assume you are a member of the United States Congress. Because of proposed legislation, you must evaluate current understanding of acid rain. A panel of scientists will testify and answer questions on the issue.

(a) You approach the panel with a completely open mind. What questions will you ask?

(b) You believe that the risks of acid rain are exaggerated. How will this change the questions that you ask?

(c) You are much concerned about acid rain and want to reduce emissions of acid precursors further. How will this affect the questions that you ask?

STRATOSPHERIC OZONE DEPLETION

In the early years of the twentieth century, highly toxic gases like ammonia and sulfur dioxide were used as coolants by America's new refrigeration industry and accidental leakage of these chemicals resulted in many human deaths. In the 1920s, the United States Congress attacked refrigerator manufacturers for producing "killer refrigerators." Then, in 1928, a young chemist announced a new coolant. This coolant could be inhaled and, when it was exhaled, the breath could be used to blow out a candle – it was not flammable. Nor was it especially toxic. This impressive coolant was a chlorofluorocarbon (CFC). Not surprisingly, CFCs were seen as a godsend and became widely used in refrigerators and, later, air conditioners. CFCs were found to have other highly useful properties. They made excellent industrial solvents, cleaning agents, and blowing agents for foam products and safe aerosol propellants. A group of related chemicals, the halons, which contain bromine rather than chlorine, became important fire-fighting chemicals. One property of CFCs

---

BOX 6.4

The stratosphere and ultraviolet radiation. The lower 10 km of our atmosphere is called the troposphere. This is the layer in which we live and in which ambient air pollution occurs. Above the troposphere, 10 to 40 km above earth, is the stratosphere. Ozone in the stratosphere performs a function vital to life on earth by absorbing much of the sun's ultraviolet (UV) radiation that would otherwise reach and damage human, animal, and plant life. Only 10% of total atmospheric ozone is in the troposphere; the other 90% is in the stratosphere, where it is formed naturally by the reaction of UV radiation with oxygen. Ozone is constantly being formed, destroyed, and reformed. A problem arises when the balance of formation and destruction is upset by anthropogenic chemicals added to those naturally present in the stratosphere.

UV radiation is divided into UV-A, UV-B, and UV-C. All of these have wavelengths shorter than that of visible light. Shorter wavelengths have more energy and thus greater potential to harm living creatures than visible light. UV-C has the shortest wavelength. UV-B has intermediate wavelength and it is UV-B that is usually referred to in discussions of UV radiation. UV-A has the longest wavelength of the three but still has greater energy than visible radiation and the potential to damage living systems.

and halons that made them so useful was their lack of chemical reactivity. In the end, this very stability was their downfall.

What are the pollutants of concern?

The CFCs and halons are the major pollutants of concern. CFCs are most commonly thought of as the coolants used in refrigerators and air conditioners. Freon (CCl_2F_2), the most widely used of the refrigerants, is employed in refrigeration systems including automobile air conditioners. Halons are used as fire-fighting chemicals. Certain *halocarbons* also pose a concern. Halocarbons are chemicals that contain carbon plus one or more halogens (fluorine, chlorine, bromine, or iodine). Other halocarbons with ozone-depleting ability are the fumigant methyl bromide, the specialized refrigerant methyl chloride, and the solvents methyl chloroform, carbon tetrachloride, and 1,1,1-trichloroethane. A 1995 report indicated that iodocarbons (iodine containing chemicals) may also deplete ozone. Also, although they do not deplete ozone in the same way as CFCs, sulfur dioxide and nitrogen oxides affect ozone.

Why is depletion of stratospheric ozone of concern?

CFCs and halons are even more environmentally persistent than polychlorinated pesticides and PCBs. Although they are volatile, CFCs are heavier than air and this fact has been used as an argument by skeptics that they cannot reach the stratosphere. However, they are so stable that they survive to mix with large air masses that move into the stratosphere and they can be detected in the stratosphere. Furthermore, unlike the highly soluble acids in acidic deposition, CFCs are not water-soluble and are not washed from the atmosphere by rain. Once they reach the stratosphere, CFCs and halons encounter more intense UV light than at the earth's surface. This radiation will eventually destroy them, but even in the stratosphere they have a long life. When they do degrade, they break down into a number of chemicals. However, the species that give rise to the reactions leading to ozone destruction are atomic chlorine and atomic bromine. If only one atom of ozone were destroyed for each atom of chlorine or bromine, there would be no problem. A problem arises because they act as catalysts – one chlorine atom destroys many thousands of ozone molecules, and one bromine atom is about 50 times more destructive still. Thus chlorine and bromine destroy ozone disproportionately to their low concentration (see Figure 6.3).

By 1976, scientists believed that CFCs threatened stratospheric ozone. At that time, two-thirds of the CFCs manufactured were used as aerosol propellants, an application banned by the United States and a number of other governments in 1979. Also in the late 1970s, plans for a fleet of supersonic jets,

Releases of CFCs, halons, and related chemicals into the atmosphere lead to

Increased stratospheric levels of chlorine and bromine, which lead to

Loss of UV-absorbing stratospheric ozone, which may lead to

Increased levels of UV at earth's surface, with likelihood of

- Increased skin cancer, eye damage, immune system suppression in humans

- Adverse effects on growth of phytoplankton

- Adverse effects on higher animals and plants

FIGURE 6.3. A stratospheric ozone-depletion snapshot.

which would have flown in the stratosphere, were dropped because of concern that the planes' nitrogen oxide emissions (injected directly into the stratosphere by their exhausts) had the potential to deplete ozone. After these actions, there was a lull in concern about stratospheric ozone. Then, in the early 1980s, British researchers began to see measurable declines in stratospheric ozone level over Antarctica in the winter months. At first they doubted their own observations. Finally in 1985, after seeing a dramatic spring decline in ozone level, they reported their findings, which others then confirmed. Antarctica is particularly vulnerable to ozone depletion because of stratospheric ice particles that form during the tremendously cold winters, particles that provide surfaces for the conversion of inactive forms of chlorine and bromine chemicals to the forms that lead to ozone destruction when the spring sun returns. Ozone levels return to normal after the weather warms and the reactions made possible by ice particles no longer occur.

Disturbing reports of ozone decline continued to emerge after the initial observations by the British team. Initially seen only over Antarctica, ozone thinning has also been reported over the Arctic during an exceptionally cold winter. Ozone losses were smaller than in Antarctica because the Arctic is less cold and forms fewer of the stratospheric ice particles that provide surfaces

for the formation of ozone-destroying species. Ozone losses have also been reported over middle latitude countries, the United States, Canada, and Europe. A United Nations Environmental Program (UNEP) report indicated
that stratospheric ozone level over midnorthern regions has decreased about
4% a decade since 1979. These losses are harder to explain than those in
Antarctica or the Arctic because there are no stratospheric ice particles to
serve as surfaces for destructive chemical reactions. Another well-known
chemical – sulfur dioxide – can explain what is happening in these warmer regions: Volcanic eruptions release huge quantities of sulfur dioxide, and a
significant amount survives to reach the stratosphere, where it is converted to
sulfate particles. Sulfate particles serve the same function as ice particles; that
is, they provide surfaces for reactions leading to chlorine-catalyzed ozone depletion.

In 1993 a NASA satellite recorded an Antarctic stratospheric ozone level of
88 Dobson units (unit of measurement for ozone) as compared to a normal
reading of about 300. This was the greatest thinning or ozone hole observed
up to that time and the effect was ascribed to sulfate particles from the Mt.
Pinatubo eruption. However, because sulfate particles survive only 2 to 3 years
in the stratosphere, investigators predicted they should be gone by 1994. Consistent with this hypothesis, ozone levels recovered in 1994 to the slower depletion rate seen before the volcano erupted. More generally, according to a
study reported in 1996, volcanic eruptions may have a major effect in the year-
to-year fluctuation in the severity of stratospheric ozone depletion. Satellite
observations indicate that, worldwide, there has been about a 2%-3% decrease
in stratospheric ozone.

Without stratospheric ozone to absorb UV radiation, life as we know it
would not be possible. All plants and animals on land and in water can be adversely affected by UV radiation. This includes phytoplankton (microscopic
algae) at the bottom of the food chain. All animals, including humans, directly or indirectly depend on plankton for food, making the productivity of
these tiny plants critical. This is one reason that a threat to stratospheric
ozone raises serious concerns. A 1985 report projected that, if CFC use continued at the 1985 rate, a 5%-7% loss of stratospheric ozone would occur by
the year 2050. This relatively small loss could increase the UV radiation reaching earth and, potentially, harm the vital phytoplankton.

It is difficult to demonstrate that lowered stratospheric ozone levels have
led to increased ground level UV radiation. This is because atmospheric conditions affect how much UV radiation reaches earth. Ground level ozone and
other air pollutants absorb part of the UV radiation. Clouds absorb some but
allow up to 80% to pass through, and they also scatter some radiation. Despite
these difficulties, separate studies in Texas, Switzerland, and Canada report
that a greater amount of UV radiation is reaching earth than in the past. A
1994 UNEP report stated that increased UV-B radiation is reaching earth's

surface, adding that the increased radiation is causing phytoplankton losses in the waters of Antarctica plus developmental damage to fish, shrimp, crabs, amphibians, and other animals. However, because of very large uncertainties in the data, quantitative estimates of adverse effects were impossible. The UNEP report warned that, in addition to living organisms, excess UV radiation may damage human-made materials, such as those plastics that are susceptible to photodegradation.

There is no "normal" level of UV radiation that reaches the earth's surface because the amount varies with latitude, time of day, and time of year. Overexposure to the sun's radiation has serious human health implications. In eyes, cataracts can result. Adverse skin effects include sunburns, premature skin aging, and skin cancer. It has been often stated that there is no such thing as a healthy tan. If UV light at natural levels can have these deleterious effects, then an increase in UV radiation resulting from stratospheric ozone depletion must be taken seriously. For each 1% increase in UV-B radiation reaching the earth, a projected 2% increase in nonmelanoma skin cancers (those associated with cumulative exposure to sunlight over the years) will occur. A more serious skin cancer, malignant melanoma, is associated, not with cumulative exposure, but with periods of intense exposure or sunburns that occur early in a person's life. Incidence of these cancers has increased rapidly in recent decades. However, skin cancer rates began their sharp increase years before stratospheric ozone declined, so changes in life-style are the suspected cause. People spend more time in the sun, often around midday, when UV radiation is most intense. They often wear clothing that is inadequate to protect the skin and 40% wear inadequate sunscreen protection. Diet may also be important to the development of skin cancer. When people who had already suffered one non-melanoma skin cancer were placed on a low-fat diet, one containing less than 20% fat, they had only one-third as many new sun-induced skin lesions and tumors as people on a typical high-fat diet had. Even though life-style is now considered the major factor in these skin cancer increases, increasing radiation resulting from ozone depletion has the potential to increase it further.

UV radiation can also suppress the immune system by damaging the skin layer, which contains immune cells. Suppression is sometimes observed as an increase in cold sores during the first sunny days of summer resulting from activation of latent (dormant) herpes virus. More generally, an increase in infectious illnesses is seen in those with suppressed immune systems; in people whose immune systems are already weakened by other illnesses or age, risk is greater. This immune system suppression not only occurs in fair-skinned people, but in the dark-skinned as well. Increasing radiation resulting from ozone depletion has the potential of further increasing these problems.

What are the sources of ozone-depleting substances?

Industry produced CFCs for more than 60 years. The United States banned CFC use in aerosol propellants in 1979, but many other countries allowed it until recent years. CFCs also served as highly useful industrial solvents, cleaning agents, and chemicals used to blow foams for cushions or polystyrene cups. The bromine-containing halons are fire-fighting chemicals. In all these applications, CFCs and halons were eventually released to the environment. The only known natural source of CFCs is volcano vents, which release – among many other chemicals – small amounts of CFCs. However, human beings generate the lion's share.

The chemical methyl bromide is a soil fumigant believed to be the largest source of ozone-depleting bromine, but a 1991 report indicated that much methyl bromide arises from marine organisms and significant amounts from forest or grass fires. Methyl chloride is also an ozone-depleting chemical, but marine and other organisms are estimated to release 5 million tons of this chemical, as compared to only 26,000 tons from human activities. Iodine, an element related to chlorine and bromine, may also play a role in ozone depletion. However, oceans naturally produce iodocarbons in amounts that dwarf human contributions and the amount of iodocarbons that reaches the stratosphere is unknown.

How are ozone-depleting substances being banned?

Although the United States banned their use as aerosol propellants in 1979, CFCs were used in increasing amounts for other purposes. Once stratospheric ozone depletion was observed, many believed that a quick and complete ban was necessary to avert serious consequences, especially because CFCs and halons have atmospheric lifetimes of decades to centuries. After a 1987 meeting, most industrialized nations signed the Montreal Protocol, an agreement to end the manufacture of CFCs and halons, and many other nations later signed the treaty, which was further strengthened in 1992. This agreement was significant because it banned ozone-depleting chemicals and because it represented the first global treaty related to environmental protection. In the United States, the 1990 CAA amendments provided for the phasing out of stratospheric ozone-depleting substances by 1996. By 1995, almost all American manufacture of CFCs and halons had ceased. In tiny amounts, CFCs will continue to be used for purposes deemed essential, as propellants in aerosol sprays used by asthmatics and in the manufacture of rocket motors. Although no longer manufactured in the United States, CFCs are still present in refrigerators and air conditioners produced prior to the ban. As these appliances reach the end of their lives, CFCs will be collected by trained technicians to prevent their escape to the atmosphere.

The major substitutes for CFCs developed to date are hydrofluorocarbons (HFCs) and hydrochlorofluorocarbons (HCFCs). But, because these alternatives have some ozone-depleting ability (about 10% as much as CFCs), better substitutes are being sought. It appears that production of these substitute coolants will be significantly lower than CFC production because leak prevention will be better and they will also be recycled. This course represents a significant change in society's approach to the handling of problem chemicals. Chemical substitutes have also been found for many industrial halocarbon solvents. But many farmers are unhappy with the alternatives available for the fumigant methyl bromide. Because methyl bromide is also produced by the ocean and by forest and grass fires, some argue that the amount used by farmers is not great enough to warrant banning its use. Although good alternatives to the fire-fighting halons are not yet available, manufacture of halons has ceased, but a 30- to 50-year supply remains on hand. Other fire-fighting substitutes that have been evaluated either have undesirable properties or are very costly. For example, one iodine- and fluorine-containing chemical showed good fire-fighting properties and had a short atmospheric life, but it was too toxic for routine use. So halon alternatives are still being actively sought.

There are obstacles to the complete elimination of ozone-depleting substances. Although production has ceased in the United States and other Western countries, the Montreal Accord allowed the eight developing countries that produce CFCs a grace period of 10 years; they need not begin to phase out production until 1999, and production need not totally cease until 2010. The reason for the grace period is the necessity for new equipment for the substitute refrigerants, which poses a problem to poor countries. In the meantime, some of these countries have increased CFC production. CFC smuggling has also become a serious problem here and in other Western countries, where the contraband CFCs are used to repair auto air conditioners. They are referred to as the second most lucrative commodity smuggled through Miami; in value, CFC smuggling is exceeded only by cocaine smuggling. So, although the world has a landmark treaty and the United States has eliminated production, much work still lies ahead.

Controlling exposure to UV radiation. Regardless of whether more UV radiation is reaching earth, it is important to control UV exposure. Children, in particular, should have sun protection because, with their outdoor habits, most have been exposed to 80% of their lifetime dose of UV radiation by the age of 18. However, the chronic effects of severe sunburn, or of overexposure even without sunburn, may not appear until many years later. For Northern

TABLE 6.2 Weather Service
ultraviolet index

INDEX VALUE	EXPOSURE LEVEL
0-2	Minimal
3-4	Low
5-6	Moderate
7-8	High
9-10 plus	Very high

Hemisphere people, there is a progressively greater need for eye and skin protection the farther south we live because the intensity of UV radiation increases closer to the equator. With all its negatives, sunlight should not be entirely avoided and is necessary to form vitamin D in the skin. Individuals who live in northern climates should be exposed to sunlight for a few minutes a day several days a week. Many people also have an improved sense of well-being with some sun exposure. However, protection is needed for anything beyond low exposures.

In 1994, the United States National Weather Service and EPA began a program to provide a summer ultraviolet index to 58 cities (see Table 6.2). The index is given to newspapers, radio, and TV as part of the regular weather forecasts. Information used to calculate the index comes from both satellite and ground-based measurements. This information can be useful to prevent excess sun exposure, which is important regardless of the status of stratospheric ozone depletion. The index is calculated for noon, or 1:00 p.m. daylight savings time. At 9:00 a.m. or 3:00 p.m., sunlight is only about half as intense as it is at noon. The higher the index value, the more quickly sunburns can occur. The farther south a person lives, the higher is the average index value. Light-skinned people or those who sunburn easily need to be especially vigilant. A high index should alert people to protect skin by wearing wide-brimmed hats and skin covering and to protect eyes with sunglasses that block 99%-100% of UV radiation. Not all skin covering is equally effective: Simple polo or T-shirts provide little protection; an unbleached cotton high-luster polyester or dark material is needed for good sun protection. There are also fabrics marketed to those with sun-sensitive skin. A sunscreen with a sun protection factor (SPF) of 15 or higher should be used. To protect against UV-A as well as UV-B, the suncreen should have an ingredient such as avobenzone added specifically for this purpose.

BOX 6.5

Problems with CFC alternatives. The chemicals used as substitutes for CFCs were selected on the basis of effectiveness in serving the desired function, lacking ozone-depleting ability, and raising no new environmental problems. For some uses of CFCs, such as metal cleaning, it was fairly easy to find substitutes; it was more difficult to replace CFCs as refrigerants or in foam used to insulate refrigerator walls. The HFCs and HCFCs that were finally selected are closely related to CFCs but have much less ozone-depleting capability. Almost all scientists believe that banning CFCs was important. However, it is important to remember that when one chemical is substituted for another, the replacement will necessarily have some environmental or safety impacts. Examples follow.

- HCFCs degrade more easily than CFCs but do reach the stratosphere and have some ozone-depleting potential. Their manufacture will be phased out early this coming century.
- HFCs have no chlorine and no ozone-depleting potential, but they are greenhouse gases and may be eventually banned for this reason.
- The hydrocarbon cyclopentane can substitute for CFCs in the production of polyurethane foam insulation, but it is a VOC, a smog-producing pollutant regulated by the CAA.
- Propellants now used to replace CFCs in aerosol cans are flammable hydrocarbons, such as propane and butane. These aerosols should not be used near flames.
- In leather shoe sprays, the ozone-depleting chemical 1,1,1-trichloromethane was replaced by hexane and 2,2,4-trimethylpentane, but incidents of acute respiratory illnesses have since occurred among those using the sprays in poorly ventilated areas.
- Water-based solvents replaced CFCs in many cleaning jobs, but the water necessarily becomes contaminated. Although they are treated before release, there is inevitably some pollutant released to water.
- Some hydrocarbons can be used as refrigerants, but they are flammable. Engineering precautions can minimize fire risk and hydrocarbons are likely to be used in Europe. However, American manufacturers fear using them because, if accidents occur, they may be sued.

Innovative substitutes are being explored. Sound waves or thermoacoustic refrigeration could be used to replace chemical coolants totally because the process needs neither refrigerants nor compressors. It combines evaporative cooling, already used in dry climates, with a desiccant to dry the air so the system can be used in moist climates. Most people would not even consider eliminating refrigeration. As one public health scientist noted, "Refrigeration has done more to increase the life span of humans than pharmacology." People used to die of food poisoning caused by microorganisms that were able to multiply rapidly in foods kept at room temperature.

Contrary views on stratospheric ozone depletion

Most scientists agree that CFCs, halons, and related chemicals present a danger to stratospheric ozone. However, some social commentators call it a scam perpetrated by scientists attempting to promote funding for their research. There is also a moderate group of skeptics who doubt that stratospheric ozone is thinning. At least some skeptics acknowledge that CFCs reach the stratosphere and some also believe that anthropogenic, not natural, sources are the major contributors to stratospheric chlorine. However, they do not believe that chlorine destroys a significant amount of ozone. Some asserted that a drop in Antarctic ozone was reported in the 1950s, before CFCs were widely used. However, the data on which that 1950s report was based were reanalyzed by a NASA researcher who concluded that instrumental errors explain what was thought in 1958 to be a lowered level of stratospheric ozone. In a 1994 *Science* article he stated, "There is no credible evidence for an ozone hole in 1958."

BOX 6.6

The final word? Some skeptics believe that the chlorine found in the stratosphere arises from natural sources such as volcanoes. If chlorine from volcanic eruptions did reach the stratosphere, the amounts would dwarf those from human activities. However, a 1993 study reported that rain washes chlorine from the atmosphere before it can reach the stratosphere and that, for example, less than 1% of the chlorine from the Mt. Pinatubo eruption reached the stratosphere. Natural sources are estimated to account for 20% of stratospheric chlorine and human activities for 80%. A NASA study reported in 1994 even more definitively concluded that stratospheric chlorine comes from CFCs as they degrade. They reached this conclusion in the following way: Satellite instruments detected the refrigerant CFC-12 in the stratosphere. The amount of CFC-12 decreased above 20 km, a level where the sun's high-energy UV radiation breaks it down. At levels where CFC-12 was detected, its breakdown products – hydrogen chloride and hydrogen fluoride – were also detected. Stratospheric levels of hydrogen fluoride have increased steadily over the years, and a steady buildup is not consistent with the hypothesis that hydrogen fluoride is produced by intermittent volcanic eruptions. Scientists directing the NASA project stated that CFCs are the source of chlorine in the stratosphere, stressing, "There is no other possibility."

Another point made by skeptics is that stratospheric ozone levels, even in one location, can naturally vary by 40% over a period of a few weeks. They further note that we have been measuring stratospheric ozone levels for less than 40 years, not long enough to know what natural fluctuations may occur. They argue that the Montreal Protocol was signed when our knowledge was too uncertain to clearly describe what was happening, let alone what might happen in the future. Critics also point out that the amount of UV radiation reaching earth varies greatly, and naturally, with degrees of latitude. Compare the UV radiation between the poles and the equator: Averaged over the year, the level at the equator is 50-fold greater than at the poles. Averaged in the summer at midday, the equator still has 10-fold greater UV radiation than the poles. For every 60 miles traveled south, UV exposure increases by 5%. Increase in elevation also increases exposure – for every additional 150 feet of elevation, UV exposure increases by 1%. For example, citizens of Denver have a 35% greater exposure to UV radiation than citizens of Philadelphia. Or, a move from Chicago south to Atlanta results in enough increased UV radiation to approximately double skin cancer risk. Despite this, most people who move south do not greatly concern themselves about increased UV exposure except that many use greater sun protection. A counterargument to this point is that life developed means to protect itself from the equator's heavier UV radiation. Consider the dark skin of tropical peoples, which partially protects them from intense UV radiation; in plant and animal life in more temperate locales similar protection has not evolved.

The future

Together, CFC-11 and CFC-12 represent 50% of the CFCs in use. Stratospheric levels of these two chemicals were increasing very rapidly until 1988, but between 1988 and 1995 the rate of increase slowed to almost zero. The data supporting this conclusion were gathered from monitoring stations around the globe, including the South Pole and Point Barrow, Alaska. However, because the lifetime of CFC-11 is 40 years and that of CFC-12 is 140 years, levels are expected only to decline slowly. Another ozone-depleting substance regulated under the Montreal Protocol is methyl chloroform. In 1995, the journal *Science* published an article reporting that the atmospheric concentration of this gas has begun to decrease – the first regulated ozone-depleting chemical to do so. This decrease is occurring because methyl chloroform's atmospheric lifetime is much shorter than those of CFC-11 and CFC-12. However, if the worldwide elimination of CFCs continues, CFC levels are also expected to fall and, as of 1996, there were indications that the decline may have begun, at least in the lower atmosphere. The stratospheric ozone level is expected to return to normal by about the year 2050.

DISCUSSION QUESTIONS

1. Assume you are a member of the United States Congress. Because of proposed legislation, you must evaluate current understanding of stratospheric ozone depletion. A panel of scientists is to testify and answer questions on the issue.
   (a) You approach the panel with a completely open mind. What questions will you ask?
   (b) You believe that the risks of stratospheric ozone depletion are exaggerated. How will this affect the questions that you ask?
   (c) You are very concerned about ozone depletion and want to make sure the provisions of the Montreal Protocol are being followed. How will this influence the questions that you ask?

2. The international treaty to eliminate CFCs obligates the world's rich nations to help poor nations with the technology they need to comply with treaty provisions.
   (a) Is this a reasonable thing to do? Why?
   (b) Many third world countries have difficulties feeding their populations and providing them with clean water or basic medical care. Under these circumstances, air pollution becomes a much lower concern. Should we be helping these nations to reduce air pollution? If your answer is yes, with what problems should we be helping? Explain.

GLOBAL CLIMATE CHANGE

Changes in climate have long been a characteristic of this planet. About 18,000 years ago, earth was in the midst of the last of many ice ages, from which it emerged about 10,000 years ago. More recently, Europe suffered a little ice age between the years 1500 and 1850. So climate change is not new. The issue we now face is whether human activities are also affecting climate through our enhanced emissions of heat-trapping gases, greenhouse gases. The warming that results from greenhouse gases has long served earth well. Radiation from the sun reaches and warms the earth's surface. In turn, earth radiates infrared radiation (IR) back toward space, but a portion is captured by water vapor and greenhouse gases. Without such a means to trap this warmth, the earth would be about 35 °C colder than it is now, too cold to support life.

What are the pollutants of concern?

Greenhouse gases are minor atmospheric components. Even the most abundant is only 360 ppm (0.036%) by volume of atmospheric gases. Compare

0.036% to the 78.1% and 21% contributed by, respectively, nitrogen and oxygen.

- Disregarding water vapor, carbon dioxide is the major greenhouse gas, accounting for about half of the total warming effect. It accounts for half, not because it powerfully absorbs IR, but because, even at 360 ppm, it is present at a much higher atmospheric concentration than any other greenhouse gas and it has a long atmospheric lifetime, 50–200 years. Until recent years, no one considered carbon dioxide in outside air a pollutant. It is a vital gas, breathed out by all animals as a waste gas and captured by earth's plants as the source of carbon used to make carbohydrates, proteins, and lipids. Essentially all biochemicals on earth result from the plant world's ability to fix atmospheric carbon dioxide.

- Second in importance, accounting for about 20% of the greenhouse effect, is methane. Its atmospheric concentration is only 1,720 ppb, a level more than 200 times lower than that of carbon dioxide. However, molecule for molecule, methane has about 20 times greater ability to absorb IR than carbon dioxide although its atmospheric lifetime of about 10 years is much shorter.

Increased fossil fuel combustion, agricultural activity, and deforestation lead to

    Increased levels of atmospheric greenhouse gases, which

        Trap a portion of IR radiated from earth, which

            Increases earth's surface temperature, with potential for

                • Increased weather variability

                • Rise in sea level

                • Spread of disease

                • Enhanced air pollution

                • Disruption of agriculture (Some positive effects also possible)

FIGURE 6.4. A greenhouse snapshot.

- Nitrous oxide accounts for about 5% of the greenhouse effect. Its concentration is 310 ppb, but it has 200 times greater ability to absorb IR than carbon dioxide. It has a long atmospheric life of at least 120 years.
- CFCs may account for as much as 20% of the greenhouse effect except that, as of 1996, CFC levels may have begun a slow decline. Also, as will be noted later, other effects of CFCs may at least partially cancel out their warming potential.
- Water vapor is the predominant heat-trapping chemical in the atmosphere, but its atmospheric concentration is highly variable and discussions of greenhouse gases generally disregard it. Neither nitrogen nor oxygen, the two gases present in the atmosphere at the greatest concentrations, absorbs IR.

Why are greenhouse gases of concern?

Four of the reasons to suspect that global warming will occur, and may already have begun, are discussed here. Figure 6.4 provides a snapshot summary of the greenhouse effect and concerns relating to it.

Atmospheric levels of greenhouse gases have increased over their historic levels and they continue to increase. At 360 ppm, the atmospheric concentration of carbon dioxide is about 25% higher than its mid-1800s level of 280 ppm. Just between 1958 and the late 1980s, it increased about 30 ppm (see Figure 6.5). Sometime between the year 2050 and 2100, its level is expected to have doubled its mid-1800s value. Methane's atmospheric level, determined by analyzing gas bubbles in ice cores, was about 700 ppb until the midseventeenth century, when it began to increase to its current level of 1,720 ppb. Nitrous oxide historically had a concentration of about 275 ppb, which began to climb early in the twentieth century to its current level of 310 ppb.

The earth has warmed in this past century. The temperature measurements leading to this conclusion were made in spotty locations around the globe, and there are also problems in interpreting the data; for instance, several different temperature-measuring devices have been used. Another major complication in trying to determine whether there has been a real warming is that many recording stations were placed in locales where cities subsequently grew up. A city, with its paving and buildings, acts as an heat island, absorbing more heat than a rural area, sometimes dramatically more. Thus, higher temperatures would be expected in these locations as compared to those a hundred years ago. However, despite difficulties in interpretation, the results of a number of research groups consistently point to a 0.3°–0.6 °C increase in temperature this past century. The critical question now is, What will happen in the coming 50 to 100 years as levels of greenhouse gases continue to increase? The most recent

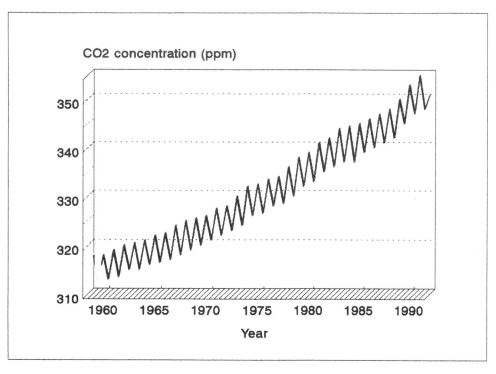

FIGURE 6.5. Increasing carbon dioxide concentrations versus time. Measurements were made at the Mauna Loa Observatory, Hawaii. *Source:* Redrawn from C. D. Keeling, R. B. Bacastow, A. F. Carter, S. C. Piper, T. P. Whorf, M. Heimann, W. G. Mook, and H. Roelloffzen, in *Aspects of Climate Variability in the Pacific and the Western Americas*, Geophysical Monograph, David Peterson (Ed.), American Geophysical Union, vol. 55, 1989.

climate models predict that earth's average temperature will increase between 1° and 3.5 °C. This may not seem a big change, but consider that the average temperature of earth's surface is only 15 °C (59 °F). During the last major ice age, which ended about 10,000 years ago, temperatures were only about 5 °C colder than today. Since then, temperature has varied only about 2 °C.

An increase of 1 °C is considered manageable. However, an increase of several degrees could be devastating, especially if it occurred rapidly with little or no time for adaptation. In that case, forests might be trapped in the wrong climatic zone and could die. Changed climatic zones could also lead to major dislocations in agriculture. The shift in precipitation patterns predicted by climate models could lead to more drought, floods, and storms, all of increased severity. Higher temperatures would also worsen air quality. Recall that photochemical smog increases on hot summer days. Another special concern is the spread of diseases. Malaria is a serious disease spread by a mosquito vector that already infects many millions in hot climates. If temperature increases occur in currently temperate regions, the mosquito vector is expected to move north, spreading malaria as it goes. A number of other diseases now restricted to hotter climates could similarly increase.

Ice core studies have led to the conclusion that there is an association be-tween high carbon dioxide levels and global warming. Analysis of air bub-bles trapped in Arctic ice cores dating back 160,000 years indicates that glacial periods were associated with low atmospheric carbon dioxide levels and inter-glacial periods with higher levels. However, whether the carbon dioxide level increased before each interglacial warming – and therefore whether it caused the warming – is not known. The level of carbon dioxide now in the atmo-sphere is described as the highest in 160,000 years.

Sea levels have gone up and ice is melting. Sea levels have increased in this past century. Whether this increase is due to warming or to natural land move-ment is unknown. However, if temperatures continue to increase, sea levels will continue to rise because water expands in response to warmer tempera-tures. This would cause coastal flooding, especially in low-lying countries such as Bangladesh and the Netherlands, and people could be driven from these lands. At the higher end of the possible warming, ice sheets in Greenland and Antarctica could melt, leading to yet greater sea level increases. Whether as a result of greenhouse gases or natural changes, Antarctica has already been warming this past century. A map drawn 100 years ago showed an island con-nected to Antarctica by an ice shelf. A modern satellite image shows the same island surrounded by water. Other NASA satellite observations show sea ice decreasing in both the Arctic and Antarctic. There has also been accelerated melting of alpine glaciers, and every tropical glacier on which there are data, has been retreating. In Venezuela, three glaciers have disappeared since 1972.

Modeling Climate Change

Predictions as to what we can expect in a future greenhouse climate come from computer models called general circulation models (GCMs). These very complex models incorporate information on the characteristics of green-house gases, their sources and environmental sinks, atmospheric circulation, and ocean circulation. There are a great many other variables. One is the role of atmospheric particulates, including the very important sulfate aerosols, which reflect part of the incoming sunlight back into space and thus have a cooling effect. Among important uncertainties in climate modeling is the ef-fect of clouds. High clouds appear to enhance the greenhouse effect, but low clouds reflect incoming sunlight back into space and thus have a cooling ef-fect. The net effect of clouds is now a cooling one. However, as greenhouse gases continue to accumulate, the effect may be either cooling or warming. Among the limitations of computer models is that they are only as accurate as the information that is incorporated in them. They can also never be com-plete; as new information is gathered, models must be continually modified.

Another limitation is the computers themselves: Although powerful, they cannot predict what will happen in any particular small region; for example, several states in the northeastern United States and several eastern Canadian provinces have been cooling in recent years. This finding does not necessarily contradict the models, which provide estimates only of what will happen over a large area of earth's surface. Because information in the models is incomplete, the predictions they produce must be interpreted cautiously and taken as best current predictions, not as absolutes.

In practice, a GCM takes a specified set of conditions and predicts global temperatures under the conditions specified. It also makes predictions on many other factors – rainfall and snowfall patterns, storm severity, and sea level. Models have made some successful predictions. They were able to predict the cooling effect on the climate of the Mt. Pinatubo eruption, which injected huge amounts of sulfur dioxide and particulates into the atmosphere. They have also had some success in reproducing the global warming that has

BOX 6.7

Volcanoes and sunshine. Atmospheric levels of greenhouse gases have steadily increased for many years. However, in 1991, the buildup of carbon dioxide, methane, and nitrous oxide slowed or stopped. Meanwhile the behavior of two other gases that do not contribute to the greenhouse effect also changed, with oxygen levels increasing and carbon monoxide levels decreasing. Scientists wondered whether there was a common cause of all these changes. The enormous eruption of Mt. Pinatubo in June 1991, which took place just before the observed changes, was suspected to be that common cause. Predictions were made that, as volcanic debris settled from the atmosphere, greenhouse gases would begin rising again. Levels of carbon dioxide and, to a lesser extent, methane did indeed begin to rise again. However, researchers are still struggling to understand why the changes took place at all. As described earlier, the Mt. Pinatubo eruption was probably also responsible for a record low stratospheric ozone level in 1992 and 1993. Variation in the output of the sun's radiation is another factor that may affect climate. Researchers have been examining a possible correlation between the sun's 11-year sunspot cycle and earth's temperature. For example, the little ice age that lasted until 1850 occurred when solar activity was very low. More recently, 1958 through 1995, earth's temperature cycle has matched the sunspot cycle, but 37 years is too short a period in which to reach definite conclusions. Nonetheless, belief in a solar effect on climate has been growing. Some calculate that about half of this past century's warming could be accounted for by variations in the sun's energy output.

taken place this past century. It is these GCMs that predict 1° to 3.5 °C of warming in the coming century and also predict the climatic effects of that warming. There are several major models and they all predict that once warming has actually begun, it will continue for hundreds of years. This is one important reason that global warming is taken so seriously: Not only does it have potentially very serious effects, but the effects would likely be long-term.

What are the sources of greenhouse gases?

Carbon dioxide. With the exception of water vapor, carbon dioxide is the major greenhouse gas. Human activities emit more than 6 billion tons a year of this gas into the atmosphere. Combustion of fossil fuel – coal, petroleum, and natural gas – is the major source. Among these, coal has the greatest carbon content and emits the most carbon dioxide when burned. The potential of fossil fuel combustion to lead to global warming was first pointed out by the Swedish chemist Svante Arrhenius in 1890. Today, major amounts of fossil fuels – probably much greater than Arrhenius could have anticipated – are burned in electric power plants and other industrial facilities. In the United States, gasoline combustion in cars and light trucks is also a significant source, emitting 20% to 25% of this country's total. Deforestation also makes a major contribution to anthropogenic carbon dioxide: When trees are burned or when they decay after being felled, the carbon stored in them is released as carbon dioxide to the atmosphere. At the same time, deforestation leaves fewer trees to absorb atmospheric carbon dioxide. Natural sources include microorganisms that respire carbon dioxide as they decompose dead plant and animal matter. Animal respiration also releases carbon dioxide, and volcanoes are a large natural source.

Most carbon is not in the atmosphere but stored, in soils, rocks, oceans, and vegetation, especially trees. Carbon exchanges among these sources and the atmosphere occur for many reasons, not all well understood. Earth's soils and oceans absorb perhaps half of the excess carbon dioxide emitted by human activities and the rest enters the atmosphere. Trees that are grown on a sustainable basis make no net carbon dioxide contribution. In *sustainable growth* as much tree biomass is grown as is harvested on an ongoing long-term basis.

Methane. A major anthropogenic source of the simple hydrocarbon gas methane is agriculture. Domestic ruminant animals such as cattle produce about 15% of all methane emissions. Rice paddies are another agricultural source of methane, which is produced there because anaerobic bacteria (bacteria that do not need oxygen), living in paddy water, break down organic material to produce methane. A nonagricultural source is landfills, where anaerobic bacteria produce methane while degrading organic wastes, food, paper, wood. and plant debris. American landfills emit an estimated 7% of the world's methane. Other anthropogenic sources are methane leaks during coal mining and flaring of natural gas from oil wells. Natural sources of methane include Arctic tundra and wetlands where anaerobic bacteria break down organic material. Methane also escapes from oil and coal deposits. Termites release methane as the result of a symbiotic relationship they have with microorganisms. So do

BOX 6.8

Is the world warming now? In 140 years of temperature measurements at the earth's surface, temperature increased noticeably during two periods. One warming period started in 1920 and continued for about 20 years before reaching a plateau. The second started in 1979, but whether it is still continuing is a source of disagreement. There are two temperature records for 1979 to 1994. One, a satellite record, shows global temperature changing hardly at all over this 15-year period. The other record, based on measurements at the earth's surface, shows a temperature increase that continued until Mt. Pinatubo erupted in 1991. The eruption injected large quantities of particles into the atmosphere, partially blocking solar radiation and cooling the earth. By 1994, most volcanic debris had fallen to earth, and, at that time, warming resumed. Both satellite and surface measurements represent good records; the problem lies in interpretation. Dr. Thomas Karl of the National Climatic Data Center in North Carolina noted that, whatever the record used, 15 years is too short a time to reach any conclusions as to whether earth is warming.

Climate models predict that, all else being equal, warming should now be averaging about 0.25 °C per decade, three times greater than the 0.09 °C per decade that has actually been observed. However, sulfur dioxide emissions are believed to be responsible for the limited warming. Burning fossil fuels releases large amounts of sulfur dioxide into the atmosphere and, periodically, so do volcanoes. Part is converted into sulfate aerosols, which create a haze that reflects the sun's incoming radiation back into space and cool the earth. Because of sulfur dioxide emissions, the United States is calculated to be about 1 °C cooler and the whole of the Northern Hemisphere about 0.5 °C cooler than would otherwise be the case. However, we cannot depend on sulfur dioxide emissions to prevent warming. They are spread out unevenly over the globe and sulfate particles have a very brief atmospheric life compared to that of carbon dioxide. Furthermore, because of the serious pollution problems to which sulfur dioxide emissions contribute, societies are working to curb them.

millipedes, cockroaches, and scarab beetles. Insects that live in the tropics or that live indoors, for instance, cockroaches, produce especially large amounts.

Other greenhouse gases. Human activities releasing nitrous oxide are combustion, soil tilling, soil fertilization with nitrogen-containing chemicals, industrial processes, and wastewater treatment. Natural sources of nitrous oxide include soils and ocean waters, both of which contain nitrous oxide-producing bacteria. Ground level ozone is a greenhouse gas, formed from VOCs and nitrogen oxides, as described in Chapter 5. CFCs are potent greenhouse gases; however, a 1992 study concluded that their warming effect may be canceled out by the cooling they cause as they destroy stratospheric ozone: Stratospheric ozone absorbs sunlight and thus warms the stratosphere. If that ozone is destroyed, temperatures should fall.

How are emissions of greenhouse gases being reduced?

There is no United States legislation mandating reductions in greenhouse gas emissions, but there is a UN Framework Convention on Climate Change passed in 1992, at the UN Earth Summit conference in Rio de Janeiro. As of late 1995, 147 countries, including the United States, had ratified the convention, whose objective is "stabilization of greenhouse gas concentrations in the atmosphere at a level that would prevent dangerous anthropogenic interference with the climate system." A major commitment made by developed countries was to return greenhouse gas emissions to 1990 levels by the year 2000. To follow up on this commitment, President Clinton in October 1993 released the United States Climate Change Action Plan. Because fossil fuel combustion is the major source of carbon dioxide emissions, energy conservation figures importantly in the plan. Some plan elements follow:

- Because industrial electric motors use large amounts of electric energy, the United States Department of Energy (DOE) will work with motor producers and users to reduce the energy they consume.
- DOE will work with electric power plants to reduce carbon dioxide emissions. If more power plants switched to natural gas, which has the lowest carbon content of the fossil fuels, emissions would be significantly reduced.
- The United States government will provide incentives to stimulate development of energy-efficient household appliances. It will also work with businesses and states to promote energy-efficient heating and lighting in buildings.
- More commuters will be asked to carpool or use public transportation. Employers may be asked to pay nondriving commuters the cash value of paid parking spots.

- The government will stimulate efforts to plant more fast growing trees as part of a reforestation program to trap a portion of the increased carbon dioxide releases.
- Government will develop partnerships with coal mine and landfill operators and with farmers and gas companies to find means to reduce methane emissions or leaks.

The Climate Change Action Plan involves voluntary steps. Some believe we must have mandatory programs because voluntary steps are not accomplishing the stated goal of reducing carbon dioxide emissions to 1990 levels. As of 1996 – far from being reduced – emissions were 4% higher than those in 1990. It is not surprising that a World Energy Council report stated there was no realistic possibility that current policies could lead to 1990 levels by 2000. If the United States is to reduce carbon dioxide emissions to 1990 levels, additional steps must be taken. One step proposed by environmentalists is to raise corporate average fuel economy (CAFE) standards for motor vehicles from 27.5 to 45 miles per gallon (mpg) for cars and from 20 to 34 mpg for light trucks. This would not only cut carbon dioxide emissions, but improve urban air quality because of lowered emissions of smog-producing hydrocarbons. Drivers would also save money driving cars with greater fuel economy. The technology is available and some European manufacturers already market cars with fuel efficiencies higher than 45 mpg. One American program, the Clean Car Initiative, has the potential to increase gas mileage greatly within a few years (see Chapter 12).

BOX 6.9

Research on global change. The United States Congress's Office of Technology Assessment (OTA) issued a 1994 report making recommendations on improving the United States Global Change Research Program. This program has emphasized research on climate change. However, global change includes a number of serious worldwide problems, including ozone depletion, biodiversity loss, population increase, deforestation, and desertification (the increase in desert land). OTA noted that the program has focused almost exclusively on climate change and spends most of its budget on earth-monitoring satellites. OTA acknowledged the importance of these but recommended that a monitoring system be established to collect information on other critical environmental changes in the world and to plan to follow these changes over a period of many decades.

Contrary views on global warming

Most climatologists agree that global warming will occur. Their major disagreement is on how much warming there will be and when it will happen. Some do not believe that significant warming will occur. They believe that the 0.3°-0.6 °C increase in temperature observed to date, even if real, is within the realm of natural variability and may have nothing to do with increased greenhouse gases. They point out that much of the recorded temperature increase occurred between 1920 and 1940, before the beginning of the most rapid rise in atmospheric carbon dioxide began. Skeptics also note points such as the following: Carbon dioxide levels were not always higher during past interglacial periods. Sulfur dioxide counteracts the effects of greenhouse gases and there will always be emissions of this gas. If warming begins, one effect would be increased formation of water vapor and increased cloud cover. Because clouds have a net cooling effect, increased clouds would counteract warming. Some skeptics simply believe that powerful forces are pushing us toward another ice age and that no amount of greenhouse gases can prevail over those forces. Even when carbon dioxide levels in past ages have been as high as now, an ice age occurred. Greenhouse skeptics also often point out that the general circulation models that predict warming were developed with too little information. Contrary to model predictions, they believe that temperature increases will be slow and modest enough that we can adapt.

Dr. Richard Lindzen of the Massachusetts Institute of Technology is a well-known skeptic. In an article in *EPA Journal*, he wrote that there is more to consider than just increases in greenhouse gases: "The Greenhouse Effect is so powerful that the Earth wisely finds ways to cool its surface. For example, by means of air currents in cumulus clouds, storm systems, and large-scale circulations, it transports heat from regions of large Greenhouse-heat absorption (near the ground, and in tropical latitudes) to regions of much-reduced absorption (higher altitudes and latitudes), thus short-circuiting over 75% of the Greenhouse Effect."

Other skeptics argue that making major changes before we have enough information is expensive and may be counterproductive. There are counterarguments to the points raised by skeptics. One is that increased cloud cover may heighten, not lessen, warming. Other scientists note that, although we must indeed expect another ice age, such an expectation does not preclude significant warming in the coming centuries, whereas another ice age may still be thousands of year in the future.

The future

Despite uncertainties, an increasing number of nations have concluded that limitations on greenhouse gas emissions are important. At a 1995 conference

BOX 6.10

Carbon dioxide is more than just a greenhouse gas. Carbon dioxide is vital to plant and tree growth. Thus – *up to a point* – increasing atmospheric carbon dioxide level may stimulate plant and tree growth and perhaps increase crop yields. This more rapid growth could also trap a portion of the growing amounts of carbon dioxide being emitted into the atmosphere. In a study of Arctic tundra vegetation that was grown in greenhouses in an atmosphere containing 680 ppm of carbon dioxide, the vegetation grew faster than it did at 340 ppm. However, growth was enhanced for only 1 year. Other studies looking at the growth of common crops indicate that fertilizing effects of carbon dioxide may last longer, although, to realize enhanced growth, crops may need greater amounts of other nutrients: nitrogen, phosphorus, and potassium. However, we cannot assume that the increasing levels of carbon dioxide in the atmosphere will have positive or even neutral effects on plant growth; some research indicates that excess carbon dioxide may stress plants and trees as they try to respond to its fertilizing effect by growing faster even under adverse conditions. An author of a 1995 *Science* article noted: "One thing seems certain: Whether air enriched in carbon dioxide warms the globe or not, the gas will alter the growth of green plants and so act as a potent force for global change."

Complicating the potential effects of increasing carbon dioxide only are effects of the nitrogen in acid rain. In a 12-year investigation, researchers studied grasses growing in a series of plots to which they had added varying amounts of nitrogen fertilizer. The nitrogen fertilizer stimulated the growth of some grass species more than that of other species. The result was *decreased* biodiversity in fertilized plots, as the faster-growing grasses crowded out the slower. The faster-growing grasses also sometimes had less than half the ability of the slower growers to act as a sink for atmospheric carbon dioxide. The amount of nitrogen fertilizer necessary to produce these effects was no greater than that deposited by acid rain in some parts of the United States The conclusion was that the nitrogen in acid rain, far from assisting in the removal of excess carbon dioxide from the atmosphere, as had been earlier hoped, has the opposite effect.

Recall from earlier discussion that the nitrogen deposited by acid rain may adversely affect forest growth over a number of years; remember too that ground level ozone, metallic, and other air pollutants can also adversely affect plant growth. Stratospheric ozone-depleting chemicals may be having adverse effects via increased UV radiation reaching the earth, but if the worldwide treaty to eliminate these chemicals is effective, that is one risk that should decline.

in Berlin, diplomats from many of the countries that had ratified the Framework Convention on Climate Change, including the United States, were charged with the task of negotiating an agreement by 1997 to reduce carbon dioxide emissions, not just to hold them at 1990 levels. The need to reduce emissions stems from the fact that the lifetime of carbon dioxide in the atmosphere is 50 to 200 years. Thus, even to maintain carbon dioxide at levels no greater than twice those existing in 1850, emissions must be reduced. Unfortunately, a 1996 report estimates that world energy demand will increase 55% between 1993 and 2015 and that a similar increase will be seen in greenhouse gas emissions. Even in developed countries, emissions are projected to rise 32%, and, in some industrializing nations, the increase may be 150%.

The growing human population has serious implications for greenhouse gas emissions. A NAS report has stated that population growth is the major single driver of atmospheric pollution. A growing population has led to a growing use of fossil fuels and greater pressure to cut forests. These activities enhance carbon dioxide levels in the atmosphere. Increasing population also leads to more agricultural activities, more cattle, and more rice paddies. These enhance methane emissions. China and India alone have a combined population of more than 2 billion and continue to grow. At the same time, these countries are increasing their use of coal as part of their programs to industrialize. If they even partially attain their goal of doubling per capita energy consumption, they will overwhelm any Western efforts to reduce carbon dioxide emissions. But, in the 1990s, it is Americans who consume 25% of the world's energy and produce a disproportionate amount of carbon dioxide. Poor nations resent the emphasis on population and point out that industrialized nations use a great deal more of resources per person than do citizens of poorer countries. The average per capita emissions of carbon dioxide in industrialized countries are about 30 times greater than in India.

Many steps proposed to reduce greenhouse gas emissions are worth taking for reasons unrelated to potential warming. Consider that – quite aside from reducing carbon dioxide emissions – better fuel economy would lead to less motor vehicle pollution while conserving petroleum, a nonrenewable fuel that makes us dependent on the unstable Middle East. Likewise, maintaining, rather than cutting down, forests is important for many reasons in addition to the fact that forests are a carbon reservoir. Stopping the world's population growth would likewise relieve many environmental stresses in addition to limiting greenhouse gas emissions. Many other examples could be given. About 150 cities are following the maxim "Think globally, act locally." Toronto is one major city that is seriously pursuing local reductions in greenhouse gas emissions. Industry has at least two viewpoints as to what we should do about greenhouse gases. Whereas the fossil fuel industry tends to downplay global warming concerns, the insurance industry has seen huge damage claims re-

BOX 6.11

A consensus on greenhouse gases. In late 1995, the UN's Intergovernmental Panel on Climate Change put out its first reassessment of climate change since 1990. The report was written by 500 scientists and reviewed by another 500. Their consensus was that "the balance of evidence suggests that there is a discernible human influence on global climate." If atmospheric levels of carbon dioxide and other greenhouse gases continue to increase, they expect earth to warm by 1° to 3.5 °C in the coming 100 years. They also concluded that variations in neither the sun's energy output nor sulfur dioxide's cooling influence would counteract the warming. They did not believe that limiting carbon dioxide levels to 1990 level was a sufficient response to the problem. Instead, they stated that emissions need to be reduced 50% in the coming 50 years.

lated to severe storms that are occurring with increasing frequency. It is concerned about climate change and wants greenhouse gas emissions reduced.

DISCUSSION QUESTIONS

1. Assume you are a member of the United States Congress. Because of proposed legislation, you must evaluate current understanding of global climate change. A panel of scientists is going to testify and answer questions on the issue.
   (a) You approach the panel with a completely open mind. What questions will you ask?
   (b) You believe that the risks of global warming are exaggerated. How will this influence the questions that you ask?
   (c) You are much concerned about global warming and want the United States to pursue actively means to reduce carbon dioxide and methane emissions. How will this affect the questions that you ask?
2. Review the Climate Change Action Plan.
   (a) Why do you believe that it has not been effective in reducing greenhouse gas emissions?
   (b) Assume that you must design a better plan. What would be the components of your plan? Explain your reasoning.
   (c) Should your community take a role in reducing greenhouse gas emissions? If so, what role?
   (d) What steps are you willing to take as one individual to lower greenhouse gas emissions?

FURTHER READING

Acid rain

American Chemical Society. 1991. *Acid Rain, Information Pamphlet.* Washington, D. C.: ACS.

Raloff, J. 1995. When Nitrate Reigns, Air Pollution Can Damage Forests More Than Trees Reveal. *Science News,* 147(6), 90–91, Feb. 11.

Smith, W. H. 1991. Air Pollution and Forest Damage. *Chemical & Engineering News,* 69(45), 30–43, Nov. 11.

Stratospheric ozone depletion

Gray, R. H., and Cooper, K. D. 1995. Ultraviolet-B Radiation and Its Health Effects. *Health & Environment Digest,* 9(3), 22–25, July.

Parson, E. A., and Greene, O. 1995. The Complex Chemistry of the International Ozone Agreements. *Environment,* 37(2), 16–20, 35–43, Mar.

Rowland, F. S., and Molina, M. J. 1994. Ozone Depletion: 20 Years After the Alarm. *Chemical & Engineering News,* 72(33), 8–13, Aug. 15.

Strange, C. J. 1995. Thwarting Skin Cancer with Sun Sense. *FDA Consumer,* 29(6), 10–14, July-Aug.

Climate change

American Chemical Society. 1990. *Global Climate Change, Information Pamphlet.* Washington, D. C.: ACS.

Graedel, T. E., and Crutzen, P. J. 1993. *Atmospheric Change, an Earth System Perspective.* New York. W. H. Freeman.

Hileman, B. 1995. Climate Observations Substantiate Global Warming Models. *Chemical & Engineering News,* 73(48), 18–23, Nov. 27.

Kerr, R. A. 1996. A New Dawn for Sun-Climate Links. *Science,* 271, 1360–1, Mar. 8.

Monastersky, R. 1995. Pine Forest Thrives on High-CO_2 Diet. *Science News,* 148(7), 101, Aug. 12.

Muller, F. 1996. Mitigating Climate Change: The Case for Energy Taxes. *Environment,* 38(2), 12–20, 36–43.

7

WATER POLLUTION

Most cities and town evolved and continue to develop along the shores of rivers and coastal areas, and human activities have greatly affected these and other water bodies. The need to maintain clean water for both humans and animals has become a major, even a critical concern. Although there were laws governing water quality before 1972, there was no uniform national law. No state controlled water pollution very well and some states, eager to keep or attract industry, were especially lax in their control. Two very significant national laws, the 1972 Clean Water Act (CWA) passed in 1972 and the Safe Drinking Water Act (SDWA) in 1974, changed that situation by mandating that all states treat water pollution uniformly. Both of these laws have been updated over the years. This chapter examines water pollution from the perspective of the pollutants themselves, how they reach water, the water bodies affected, and the pollution of drinking water.

Water can be divided into groundwater and surface water, which include rivers, lakes and streams, wetlands, coastal waters, and oceans. Water can also be categorized as fresh water or brackish water (salty or briny water) like sea water. Marine water is not drinkable and some groundwater is also too brackish to drink. Certain other surface waters, such as the Great Salt Lake in Utah, are also salty. In addition to the usual four questions regarding pollutants – pollutants of concern, why they are of concern, and so on – an additional question arises for water: What type of water body is polluted? A given pollutant may have an adverse effect or little or no effect, depending upon the type of water body in which it is found. Runoff of soil in rainwater into a fast-running river may not pose a problem, but the same amount of soil discharged into a lake or slow-moving stream may cause severe damage. An organic pollutant discharged into surface water may be degraded by microbial action or by action of oxygen and sunlight, or it may evaporate. The same chemical in groundwater, where little means exist to degrade it, may become a long-term contaminant. Metals are natural components of marine waters and

BOX 7.1

Pollutant movement among land, water, and air. Chapters 5 through 8 examine pollution of three environmental media, air, land, and water. Discussing each medium separately is often the easiest way to analyze pollution and the way that environmental agencies categorize it. However, a specific pollutant may contaminate all three media and, sometimes, cycle among all three. Chemicals emitted into the air settle onto water and land, including on food crops and other vegetation. Likewise, pollutants discharged to water or applied to or disposed of on land can become airborne. Many land contaminants run off into surface water or seep down into groundwater. PCBs are an example of a pollutant that can move among air, water, and land. Most PCBs were discharged into water or onto land. Although not especially volatile, they can and do become airborne from both water and land. Once airborne, they eventually settle again onto land and water, often far from their origin. From these second resting places, they again become airborne, and the whole cycle continues to repeat itself for the lifetime of the PCBs. An estimated 91% of the PCBs entering Lake Superior are from the atmosphere. Acid rain is another type of example. Pollutants originally emitted only into the atmosphere land on the earth to affect both soil and water. Or, consider that 69% of the anthropogenic lead and 73% of the mercury in Lake Superior reach it by atmospheric deposition.

a small input of a metal may not be noticeable. The same amount may cause problems in freshwater bodies, where metal concentrations are normally low.

WATER POLLUTANTS

Before discussing specific water pollutants, it is important to examine several terms important to the understanding of water pollution. One of these is point source. EPA identifies a *point source* as "any single identifiable source of pollution from which pollutants are discharged, e.g., a pipe, ditch, ship, or factory smokestack." For example, the outlet pipes of an industrial facility or a municipal wastewater treatment plant are point sources. In contrast, a *nonpoint* source is more diffuse. In urban areas, nonpoint sources include city streets, parking lots, and construction sites and, in rural areas, agricultural, logging, and mining lands. Rainwater and snow melt runoff from these sources carry oil, grease, dirt, trash, animal waste, microorganisms, and chemical pollutants, including metals, pesticides, and fertilizers. Atmospheric deposition of pollutants is also a nonpoint source of acid, nutrients, metals, and other airborne pollutants. The word *runoff* signals a nonpoint source that originated on land. When land is undisturbed by human activities, rainwater is free to percolate through the soil, which absorbs and detoxifies many pollutants. But on land disturbed by parking lots, roads, shopping malls, factories, buildings, and homes, there is little soil, and contaminated rainwater runs off into area water bodies. The United States EPA considers polluted runoff the most serious of this country's water pollution problems.

Conventional pollutants

What are the pollutants of concern? Water pollutants regulated under the CWA are conventional pollutants, toxic pollutants, and nonconventional pollutants, in particular, those discharged into surface waters, rivers, lakes, wetlands, estuaries, bays, and oceans. Just as the word *criteria* in the term *criteria air pollutants* does not reveal the seriousness of criteria air pollutants, even less does the word *conventional* in conventional water pollutants tell us that these pollutants can have devastating effects. However, the use of the word does imply that these are common pollutants produced in large amounts. The six pollutants are biochemical oxygen demand, nutrients, suspended solids, pH, oil and grease, and pathogenic microorganisms. None of the conventional pollutants is a single chemical and one, microorganisms, represents living organisms. Each will be discussed separately.

Biochemical oxygen demand (BOD). The sewage and other organic matter discharged to a water body are degraded by oxygen-requiring microorganisms. The amount of oxygen consumed by the microbes is the *biochemical oxygen demand* (BOD). Although some natural BOD is almost always present, BOD is often an indication of the presence of sewage and other organic waste. High levels of BOD can deplete the oxygen in water. Unless fish and other aquatic organisms have the freedom to escape from low-oxygen (hypoxic) conditions, they may die. Professor R. Diaz of the College of William and Mary's School of Marine Sciences has noted, "Low oxygen now causes more mass fish deaths than any other single agent, including oil spills and it ranks as a leading threat to commercial fisheries and the marine environment in general." Hypoxic water is a problem in the United States and worldwide. It can be an irregular or seasonal occurrence or may last for many months. Natural sources of the organic matter exerting a BOD include dead plant and animal debris and wild animal and bird feces. Anthropogenic sources include discharges from municipal wastewater treatment plants, food-processing operations, chemical plants, pulp- and-paper-making operations, tanneries, and slaughterhouses. The synthetic nutrients in fertilizers also contribute to BOD by stimulating the growth of algae and other plants that exert BOD after their death.

Nutrients. A nutrient is a substance required for life; it promotes growth. But remember that a nutrient acquires a more sinister face at high concentrations. Synthetic fertilizers are composed of nutrients. When discharged to water, the nitrogen and phosphorus they contain become available to water organisms. Nutrients are also discharged as organic matter, which contains nutrients that become available to water organisms as the organic matter degrades. A rich supply of nutrients entering a lake, estuary, or bay may hasten *eutrophication*, which EPA defines as a process "during which a lake, estuary, or bay evolves into a bog or marsh and eventually disappears." As the water body ages, it accumulates the nutrients leading to algae growth, and, during later stages of eutrophication, the water becomes choked with plant life. Eutrophication can be entirely natural, happening slowly as a water body ages over many years, or can be accelerated by human discharges of nutrients. Runoff from farmland to which fertilizer has been applied is rich in nitrogen and phosphorus that can stimulate a *bloom* (a very abundant dense growth) of aquatic plants, especially algae.

In marine coastal water, a particularly toxic group of organisms, the dinoflagellates, can be stimulated to bloom by high nutrient levels. Their bloom is referred to as a red tide. Humans who eat shellfish that have fed on red tide organisms may suffer from paralytic shellfish poisoning caused by the toxin that they produce. This is the reason a ban is placed on eating certain shellfish in some summer months. One particularly dangerous dinoflagellate lives in warm coastal marine waters from North Carolina south and has been identified as the cause of bizarre fish kills. In 1995, huge numbers of dead fish described as having red weeping wounds were found floating in several coastal rivers of North Carolina. These wounds, actually ulcers, were attributed to ichthyotoxin, the toxin produced by this dinoflagellate (see Figure 7.1). When the algae in an ordinary algal bloom or a toxic algal bloom die, oxygen-requiring microorganisms decompose them. Thus, the algae blooms that have been stimulated by nutrient input exert BOD and can lead to the hypoxic water described above in relation to BOD. Thus, it is not surprising that damage from nutrients is described in terms similar to those used for describing the danger of hypoxic water. In 1990, a UN report referred to nutrients as the most damaging class of pollutants in the marine realm.

A 1996 *Environmental Health Perspectives* article described what can happen when there is a marked increase in the discharge of nitrogen and phosphorus into a water body, in this example, North Carolina's Pamlico River estuary.

FIGURE 7.1. Effect of a dinoflagellate toxin on fish. *Source*: E. J. Noga, DVM, North Carolina State University, 1995. Reproduced with permission.

This estuary was a rich fishing ground in 1980, but by 1996, both commercial and recreational fishing were gone. An estimated 15% of the nutrients that damaged this estuary are from point sources, the municipal sewage plants in North Carolina's fast-growing cities, and the other 85% was in runoff from

---

BOX 7.2

A huge dead zone. Scientists are only now grasping the vastness of area of oxygen-depleted – and life-depleted – bottom waters in the Gulf of Mexico near the mouth of the Mississippi River. The dead zone is attributed to nutrients in runoff from fertilizer used on farms in the Mississippi River basin. An estimated 16% of the nitrate in fertilizer applied to crops in this region runs off into the river and is carried to the mouth of the gulf. This is a massive dose of nutrients and the resulting dead zone may be one of this country's major ecological problems. Remediation could require changes in ways United States farmers fertilize in a region that constitutes almost one half of the area of the United States.

---

forestry and farming operations, livestock waste ponds, and cities. Major amounts of nutrients are from farmland to which fertilizers have been applied, from manure piles, and from animal feeding lots or lagoons containing animal wastes. Another water body receiving massive doses of nutrients is the Gulf of Mexico: An estimated 16% of all nitrate fertilizer applied to crops in the Mississippi River basin run off into the Mississippi and are subsequently delivered to the Gulf (see Box 7.2). Acid rain is another nonpoint source of nutrients to some water bodies, representing about 40% and 30%, respectively, of the nitrogen input into Chesapeake Bay and the Potomac River basins. Natural sources of nutrients are similar to BOD sources. They are present in the water's dead plant and animal debris and in wild bird and other animal feces. Another natural source is runoff from the debris on a forest floor.

Runoff from livestock operations merits further comment. Mammoth livestock farms have become increasing sources of water pollution. At any one time, the United States has 60 million hogs, 47 million beef and dairy cows, and 7.5 billion chickens. These animals, typically kept in confinement, produce about a billion tons of manure and urine a year, often stored in lagoons. In a 1995 incident in North Carolina, a lagoon leaked 25 million gallons of hog waste into the New River, leading to an algal bloom, oxygen depletion, and killing fish and other aquatic organisms. Another serious consideration is chronic seepage from lagoons containing animal waste and runoff from the fields where lagoon liquids are sprayed. Not only is surface water contaminated, so is groundwater. In Colorado towns near huge beef cattle feedlots, nitrate concentrations in groundwater are double the EPA standard. In addition to polluting water, livestock farms generate local ammonia air pollution. In North Carolina, ammonia in rain increased 25% between 1990 and 1995, coincident with the increase in hog farming.

DISCUSSION QUESTIONS

1. Dr. Stephen Eisenreich of Rutgers University recently noted that eutrophication is probably the major water quality problem that the
United States faces. He further commented that, although environmental health is critical to the ongoing economic development that
Americans desire, they are more willing to accept continuing eutrophication and other adverse environmental impacts resulting
from urban sprawl than to change land use practices.
   (a) If you believe that Americans should change land use practices,
   what effective options exist to modify behavior?
   (b) What can you as an individual do to ameliorate the ill effects of
   "sprawl"?
2. (a) Consider the information in Box 7.2. From what you understand of the effects of a nutrient overdose, explain how the Gulf
   of Mexico dead zone could have come about.
   (b) Examine paragraph two in Box 6.10. Are the effects taking
   place in this terrestrial system similar to eutrophication of water
   bodies? Explain.

pH. The term *pH* does not refer to a chemical. As described in the Chapter 6
discussion of acidic deposition, pH refers to the acidity or alkalinity of water. A
low pH may mean a water body is too acid to support life optimally. Some water bodies are naturally acidic, but others are made so by acidic deposition or
acid runoff from mines.

Suspended solids. Suspended solids are physical pollutants. They are always
naturally present in water to some extent, and, as usual, it is an excess that is
deleterious. Fine particles from soil runoff can remain suspended in water
and increase its turbidity or cloudiness. This can stunt the growth of aquatic
plants by limiting the amount of sunlight reaching them. Suspended solids
can also clog the gills of fish and other water animals and impair respiration.
Excessive turbidity also interferes with water disinfection by shielding microorganisms from the disinfectant, potentially resulting in live pathogens entering drinking water. Soil runoff from agricultural and forestry operations
and construction activities is one of the anthropogenic sources of suspended
solids. Many effluents from the same facilities that discharge BOD also contain suspended solids.

Oil and grease. Oil spills are a major problem in near-coastal waters and can
kill or adversely affect fish, other aquatic organisms, and birds and mammals.
Spills can kill or reduce populations of organisms living in coastal sands and
rocks, and may kill the worms and insects that serve as food to birds and other

> **BOX 7.3**
>
> **The *Exxon Valdez*.** In 1993 the biologist Rick Steiner described the effects of the Alaskan oil spill as follows: "The essence of the disaster lies in images of once-playful river otters oiled and crawling off to die in rock crevasses along their home streams; bald eagles losing their grip in the treetops, falling dead deep in the forest; orphaned sea otter pups searching for dead parents, shivering through oiled fur in cold water that once seemed warm; seals, sea lions and whales staring up at a black surface through which they must swim in order to take their next breath, eyes and nostrils inflamed, often then inhaling oil instead of air; diving birds, soaked in oil and unable to fly, with simply nowhere to go but back into the thick of the oil. If nothing else, the *Exxon Valdez* should serve to remind all of us that any true prosperity we seek in this world must also include consideration for the many innocent beings along the way."

animals. When oil intrudes into coastal marshes, it damages or kills fish, shrimp, birds, and other animals. An oil spill may also foul beaches used for swimming and recreation. Depending upon the amount and type of oil spilled, where it is spilled, and weather conditions, recovery can be quick or painfully slow. Despite the sometimes horrendous results of oil spills, they have overall been described as a relatively minor problem for fish and the marine environment in comparison to, for example, chronic nutrient pollution. Spills are not the only source of oil in water: Oil leaking from vehicles or released during accidents is washed off roads with rainwater and into water bodies. Improper disposal of used oil from cars is another source; motor and other recreational boats release up to 30% of their fuel, unburned, into water bodies (this source will be regulated in the near future). Altogether, these individually small, but ongoing pollution events add up to much more oil than is spilled in dramatic events such as that of the *Exxon Valdez*. However, whereas the effects of a major spill are obvious, the environmental impact of ongoing small events is harder to assess.

Bacteria, viruses, and protozoa. The concern associated with microorganisms is infectious disease. Pathogenic organisms that are contaminating offshore water can directly infect humans who come in contact with the water and cause shellfish beds to be closed to harvesting. Microorganisms are naturally found in water and elsewhere in the environment and can cause infections. However, the microbes causing the greatest concern are usually associated with human activities. Nonpoint sources include runoff from livestock operations and storm water runoff, especially that associated with combined sewer overflow (defined later). Poorly performing municipal sewage treatment plants are point sources of pathogenic microorganisms. Infectious microorganisms are a tremendous threat when found in drinking water, a subject which will be further discussed later.

Toxic pollutants

Recall from Chapter 5 that criteria air pollutants are as toxic as HAPs. Likewise, conventional water pollutants can be as harmful as toxic pollutants. As is the case with hazardous or toxic air pollutants, there are many hundreds of potentially toxic water pollutants. Of these, EPA, under the CWA, regulates 126 of special concern, the priority pollutants. These can kill or sicken fish and other aquatic animals. Inorganic pollutants among the 126 include arsenic and the metals cadmium, lead, mercury, nickel, copper, and zinc. Among the organic pollutants controlled by law are the widely used industrial chemicals benzene and toluene and many pesticides. Toxic pollutants from specific facilities can be controlled, but it is more difficult to control nonpoint sources such as air deposition or rainwater runoff. In 1992, 43 states reported a total of 930 fish kills caused by runoff and spills of pesticides and other pollutants. A number of the toxic water pollutants, including benzene, toluene and the metals, have already been discussed as toxic air pollutants in Chapter 5 (see Table 5.3 again). This overlap results from the fact that many pollutants are transmedia or crossmedia pollutants that can contaminate air, water, or soil.

Other pollutants of concern

A third category regulated under the CWA is nonconventional and nontoxic pollutants. Ammonia, chlorine, iron, and total phenols are in this category. So is color. Many facilities – textile factories are an example – discharge colored effluents. The intensity of the color that can be released is regulated by law. Another regulated pollutant is heat. Electric power plants especially, but also many industrial facilities discharge heated effluents. Thermal pollution can cause problems but is not ordinarily a serious pollutant. The discharge of certain chemicals, including radioactive chemicals and chemical and biological warfare agents, is totally forbidden by law. Another group of chemicals, whose discharge is prohibited, is the PCB family. These are regulated, not by the CWA, but by the Toxic Substances Control Act (TSCA). Manufacture of PCBs was banned in the 1970s, but they are still found in runoff entering water bodies from old spills, waste sites, or leaks in old electrical equipment. As a result of large PCB releases that occurred prior to the ban on their manufacture, there is also a reservoir of PCBs in sediments.

BOX 7.4

The contribution of sewers and storm drains to water pollution. A sewer is an underground pipe system that carries wastewater to a treatment plant.

- A sanitary sewer carries wastewater from homes and commercial and industrial establishments to a treatment plant.
- A storm sewer carries runoff from rainstorms or melting snow.
- A combined sewer carries both sanitary wastewater and storm water runoff. Sewer contents are ordinarily carried to a wastewater treatment plant, but when the plant cannot handle all the runoff reaching it during a storm, it overflows. The overflow, containing untreated sewage as well as storm water, is referred to as combined sewer overflow (CSO). Whereas the outlet pipe of a wastewater treatment plant is a point source of pollution, CSO becomes runoff that contributes to nonpoint pollution; combined sewers, which date from the late nineteenth and early twentieth century, are still used in about 1100 eastern and midwestern communities.
- Storm drains also contribute to pollution: In some cities, runoff entering a storm drain is directly discharged into nearby water without treatment. Storm drains, not ocean dumping, as previously believed, may have been responsible for most trash on U.S. beaches in the 1980s. Storm runoff, along with CSO and other poorly controlled sewage systems, led to the issuance of 1,400 beach warnings or closings in 10 coastal states in 1990.

Up to 19 out of 1,000 swimmers each year are reported to have gastroenteritis caused by swimming in water containing infectious microbes. Contamination with pathogenic microbes was the primary reason for closing beaches, but other contaminants contributed to the problem. About a third of the cases of water quality impairment nationwide are attributed to storm water, which, until recently, was the largest remaining unregulated source of water pollution. EPA has now established a two-phase program to regulate problems associated with wet weather. Large communities, those with more than 100,000 people, will be regulated first, along with large construction sites and industries. EPA has also developed a policy to control combined sewer overflow and sanitary sewer overflow. These major and expensive programs will take time to implement effectively.

How are discharges of water pollutants being reduced?

The intent of the CWA was to eliminate eventually point sources of pollution. But, because elimination was not immediately possible, it issued permits to municipal and industrial facilities that allowed discharges of specified amounts of particular pollutants. A facility has to comply with its permit and must regularly monitor its discharges to assure that it is in compliance. The goal of the permits is to protect human health and aquatic life.

Wastewater treatment. Tap water becomes wastewater after it is contaminated during use in homes and businesses for drinking, bathing, flushing toilets, and other purposes. Before a municipal or industrial facility can discharge wastewater into receiving water, wastewater must be treated to remove conventional and nonconventional pollutants to a level in compliance with its particular permit. Figure 7.2 summarizes wastewater treatment. Primary treatment first removes large objects such as sticks and trash, and then smaller objects such as sand and small stones. Suspended solids remain in the wastewater; these are settled out in a sedimentation tank. After solid materials have been removed, the wastewater moves on to secondary treatment, where microorganisms digest soluble contaminants. These steps do not remove all conventional pollutants. One major deficiency is the inadequacy of this process to remove the nutrients phosphorus and nitrogen, which can be detrimental to receiving waters. Advanced treatment must be used to remove these. As necessary, advanced treatment is also used to remove those priority pollutants whose concentration has not been adequately reduced by basic treatment. The sludge that results from wastewater treatment is often called biosolids to indicate its potential use as a fertilizer or compost. In some locales, especially those with industry, such uses of biosolids are controversial because of the contaminants they may contain. EPA regulates releases of specific toxic pollutants by requiring facilities to use so-called best available technology to limit releases to the permitted level. A less stringent best conventional technology is required for conventional pollutants.

In earlier years, industry often sent its untreated wastewater effluents directly to a municipal facility, paying it to treat them along with municipal wastewater. But industrial effluents sometimes have components that interfere with proper functioning of the municipal plant. Or the plant may be unable to remove noxious pollutants and pass them into receiving waters. Also, municipalities often want to use biosolids for beneficial purposes; and the presence of certain industrial pollutants may make these applications impossible or more difficult. So EPA began requiring industrial plants to pretreat their wastewater before releasing it to a municipal plant. Alternatively, an industry may take complete responsibility for its own wastewater treatment and directly release it to a waterway without going through a municipal plant. To

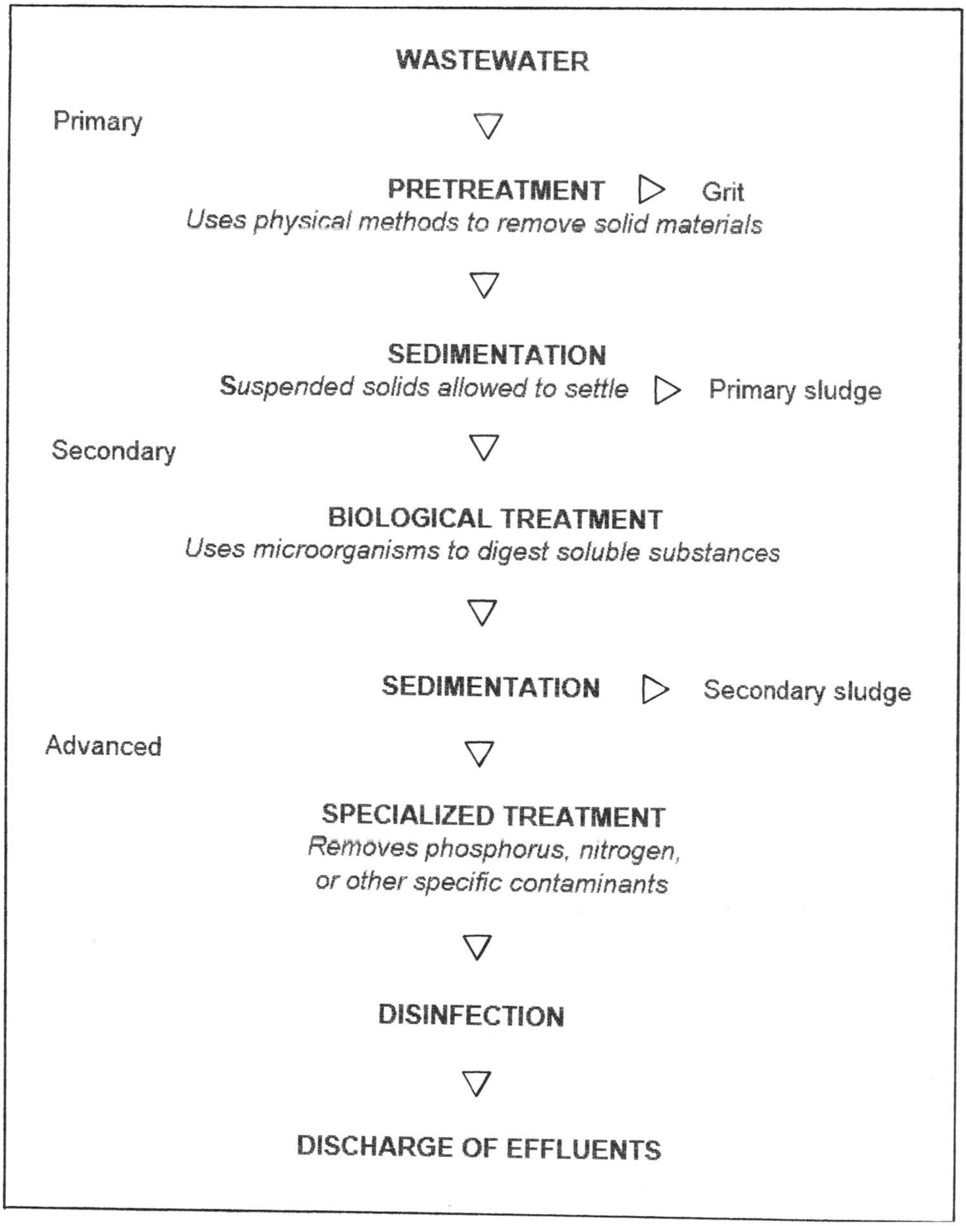

WASTEWATER

Primary   ▽

                    PRETREATMENT ▷ Grit
             Uses physical methods to remove solid materials

                                    ▽

                    SEDIMENTATION
             Suspended solids allowed to settle ▷ Primary sludge

Secondary   ▽

                    BIOLOGICAL TREATMENT
             Uses microorganisms to digest soluble substances

                                    ▽

                    SEDIMENTATION ▷ Secondary sludge

Advanced   ▽

                    SPECIALIZED TREATMENT
             *Removes phosphorus, nitrogen,
             or other specific contaminants*

                                    ▽

                    DISINFECTION

                                    ▽

                    DISCHARGE OF EFFLUENTS

FIGURE 7.2. Simplified flowchart of wastewater treatment. Advanced treatment is not always used. *Source:* Adapted from *Preventing Waterborne Disease*, U.S. EPA, Office of Research and Development, EPA/640/K-93/001.

do so, the industry often needs specialized processes to remove pollutants of special concern.

Alternative methods of treating sewage and wastewater. Millions of households and small businesses in rural areas or small communities are not connected to a public sewer line. They use on-site systems, of which septic systems

BOX 7.5

One small business's approach to water protection. Small business operations sometimes seriously contribute to local pollution. Those posing the biggest threat to a watershed are vehicle maintenance, dry cleaning, printing and photoprocessing, light construction, metal finishing, and metal fabrication businesses. Each may only generate small quantities of hazardous waste, but these businesses exist in such large numbers that the total amount is significant. This multitude of small businesses is difficult to regulate, and many of their owners lack the knowledge and resources to respond fully to requirements placed on them. However, some, like the Brooklyn General Repair shop, act in a responsible and innovative manner. This small Connecticut business reconditions radiators for cars, trucks, and industrial equipment and handles grease, oil, acids, solder, and heavy metals. Its owner researched, bought, and installed a system to handle the polluted wastewater generated by steam cleaning of vehicles and floors. Floors were sloped toward a central drain to collect water, grease, sludge, acid, and dirt. The drain was connected to a series of separators that remove grease, waste oil, and sludge. The collected waste oil is combined with heating oil and used to heat the business in winter. Sludge and grease are sent to disposal sites. The wastewater is filtered and treated to a level of purity that allows it to be reused for steam cleaning.

are the best known. These systems produce a concentrated *septage*, which must be periodically pumped out and taken to a municipal plant for treatment. In rural areas, septage is sometimes spread directly on land. Publicly owned wastewater treatment systems are very expensive, not only because of the treatment plant itself, but because the waste must be collected through a pipe system from each home and carried to a central plant. Because of the expense of installing a central municipal system, small communities may prefer homeowners to build individual on-site systems. However, improperly installed and maintained systems often fail and, in some locales, are a serious nonpoint source of pollution. Properly built and maintained onsite systems, specialists believe, can treat waste in an ecologically sound manner and return the water directly to the environment. To assure that private systems are properly maintained, one locale in Washington state issues a permit for each new system. The permit must be renewed every 4 years, and before renewal the homeowner must show that the system has been properly maintained.

Although septage from on-site systems may be taken to and treated at a traditional wastewater treatment plant, alternatives are evolving. One Massachusetts company, Ecological Engineering Associates, uses a greenhouse to treat household septage. Within the greenhouse, the septage is aerated; heated; treated by bacteria, algae, plants, and snails; and finally run through an engineered marsh. After this 8-day process, a clear, odorless water emerges that is suitable for irrigating crops, watering lawns, or discharging to rivers and lakes. In fact, the water is cleaner than that discharged into the ocean by Boston's new wastewater treatment plant. At the same time, the system uses less energy, eliminates the need for chemical additions, and is less costly to operate than a municipal wastewater treatment system. Unfortunately, such a system is not feasible for a large city like Boston. Another alternative to traditional wastewater treatment for small communities is to use artificial wetlands, so-called constructed wetlands. These absorb sediments and microorganisms from the wastewater, and wetland plants often degrade nutrients and toxicants. Even in states with cold climates, communities can use constructed wetlands by containing them in greenhouses. Although not feasible for large cities, they are viewed with increasing interest by small communities.

It is sometimes pointed out that although it is treated, there is something strange about collecting and discharging mammoth amounts of human waste into waterways. Is there a more environmentally sound way to handle it? The old outhouses have sanitary, environmental, and comfort problems, and few consider them an option. Some alternative toilet options were recently discussed in *Small Flows*, an EPA-supported publication of West Virginia University. Composting toilets and electric incinerating toilets are among these alternatives, as is one that combines the use of solar energy with composting. At this stage, alternative toilets are envisioned for rural homes and small communities. A high-technology toilet was developed by NASA researchers for use in space. This elaborate toilet cost $23 million to design and build and can only be used in space. But that it could be envisioned at all holds promise for better ways of handling our waste. Considering the huge pollution problem posed by an already large human population that continues to grow and human settlements that continue to expand, alternative ways of handling human waste demand more attention.

Reusing wastewater. Household *gray water* (all wastewater produced by a household with the exception of sewage) is now disposed of along with sewage; this method greatly increases the volume of wastewater and the cost of treating it. Collecting and reusing gray water within individual households is being taken very seriously in some parts of the United States. The recovered gray water can be used to flush toilets or, in some cases, to wash cars or water yards. Especially in locales where fresh water is limited, there is much appeal in reusing water. There is also the larger issue of wastewater reclamation (all wastewater: gray water plus sewage). Urban areas are especially interested in water reclamation and reuse because of the increasing scarcity of water in dry areas, increasing cost of treating drinking water, and increasing population. Among many possible reuses of treated wastewater are as cooling and process water and as water for commercial washing, ornamental fountains, fire fighting, irrigation of golf courses, creation of artificial wetlands, and groundwater recharge. The major pollution risk of using reclaimed wastewater arises from infectious microorganisms. But, because reclaimed wastewater is not used for drinking water, pathogens need only be reduced, not eliminated.

Successes and problems in reducing wastewater pollutants. EPA has controlled releases of conventional and toxic pollutants so successfully that, by 1994, only about 15% of water pollution was still traced to those sources. Emissions of some pollutants were reduced 90% or more. This water quality improvement occurred despite a 25% increase in population and a 50% increase in gross national product. When the CWA was passed in 1972, only 30% of American waters were judged fishable and swimmable. In 1994, this figure was greater than 60%. *Fishable* means that fish from the water are safe to eat; *swimmable,* that the water can be used for swimming without fear of infectious organisms or other contaminants at levels potentially harmful to health.

Control of point sources has been very successful, but point sources are sometimes still significant, especially in urban areas. Also, it has not been possible to meet the original intent of the CWA, which was to eliminate pollutant discharge from point sources. In fact, the name of the permit that allowed municipalities and industries to release specified amounts of pollutants was the National Pollutant Discharge Elimination System (NPDES). The word *elimination* was deliberately used because the intent was to phase out pollutant discharges entirely. Recall from Chapter 1 the difficulty of totally eliminating a pollutant by end-of-pipe means. Even with very expensive controls, total elimination can be impossible; municipalities, especially, have a difficult time finding money to modernize or even properly maintain wastewater treatment systems, some of which have deteriorated to the point that untreated sewage is sometimes released. On occasion, sewers have even overflowed from manholes into streets. In 1991, EPA estimated that modernization of these facilities would cost $110 billion by the year 2010.

Some organizations have pressured Congress to require EPA to follow through on the intent of the original CWA and set a timetable to eliminate pollutant releases. However, a spokesperson for the Water Environment Federation, an organization of scientists, engineers, and wastewater managers, observed that many pollutants exist at levels so low that they are hard to quantify and increasingly difficult to eliminate. He stated, "It would probably cost as much to eliminate the last 5% of a contaminant as to eliminate the other 95%. Our philosophy is to move toward smaller amounts of pollutants and analyze the cost of eliminating decreasing quantities versus the benefit of the environmental gain." Reconciling the desire to require pollutant elimination with the costs and difficulties of accomplishing elimination is difficult. Many environmental organizations believe zero discharge can be accomplished by reliance on P². Each pollutant would be given a deadline by which its discharge had to be eliminated. Very toxic chemicals would be phased out more rapidly than less toxic ones. Despite problems with controlling point sources of pollution, the major water quality problem remaining in the United States is nonpoint runoff.

Controlling and reducing nonpoint runoff

Runoff of polluted rainwater and snow melt is the largest remaining source of water pollution in this country. Speaking to a Senate panel in 1993, EPA administrator Carol Browner discussed water protection needs that Congress should emphasize when it reauthorized the CWA. She stated that a watershed protection approach must be taken. A watershed is a drainage basin, the area that drains all the surface water in the area into a particular stream, river, lake, aquifer, or estuary. Surface water and groundwater are closely interconnected and both can be contaminated by runoff, so a holistic approach to protect both is necessary. Browner told the panel that reducing the pollution resulting from runoff needs to be "the single greatest achievement of an amended Clean Water Act." And P² needs to be the approach most emphasized, that is, producing fewer of the pollutants that find their way into runoff.

Reducing nonpoint runoff from agriculture. Agriculture is a major source of nonpoint runoff into rivers, lakes, and other water bodies that deposits soil, pesticides, fertilizers, and animal wastes. Means to reduce runoff vary with the type of farm and local conditions. Planting a buffer strip of vegetation next to a water body to absorb runoff is one control. No-till farming is becoming an important means to limit the serious problem of soil erosion and soil runoff into water. In no-till farming, crop residues are left on the soil instead of being tilled into the ground. However, no-till farming is one more example of an environmental trade-off because it requires increased use of herbicides to manage the weeds that are not plowed into the ground. Fertilizers (nutrients) are another component of farm runoff; farm crops, for example, potatoes, corn, or hay, cannot capture all the fertilizer with which they are provided, and farmers must typically apply more than the crop actually needs. When it rains, the excess runs off into surface water or may percolate down into groundwater. In some locales, agricultural specialists are working with farmers to analyze farms for nutrient levels, section by section. Results are entered into a computer placed within a tractor. The computer information lets the farmer know whether a particular section actually needs fertilizer before any is applied. With less fertilizer applied, there is less fertilizer runoff into water.

Because pesticide runoff is such a major concern, *integrated pest management* (IPM) is increasingly used by farmers to determine when a pesticide is really needed, as contrasted to making routine applications according to a calendar schedule. Reduced applications mean less pesticide in runoff from treated fields. Chemical companies are also beginning to produce herbicides that can kill weeds in much lower amounts; again, if less is applied, less remains to run off into water. Newer herbicides are often less soluble in water and bind more tightly to soil, characteristics that also lower pesticide runoff. Runoff from animal farms often contains animal waste, which can contaminate water with potentially dangerous bacteria, viruses, or the protozoa *giardia* and *cryptosporidium*. Effective control of agricultural runoff is a major challenge because it requires farmers to change their practices greatly. They need education on methods to reduce runoff and incentives to make use of that education. An example of a P^2 project is a cooperative venture of dairy farmers in Wisconsin's Green Bay watershed, the University of Wisconsin's Cooperative Extension Service, and the USDA. The project's intent was to reduce runoff of pesticides, fertilizers, and eroded soil to waterways. Crop consultants visited participating farms each week of the growing season for 3 years to monitor pest populations in the crops being grown to feed cows with the intent of limiting pesticide applications to those crops. Farmers were also encouraged to plant nitrogen-fixing legumes to enrich the soil and thus reduce the need for fertilizer. Farmers did reduce pesticide and fertilizer use; and the result was less contaminated runoff. In addition, participating farmers saved an average of $5,000.00 a year, $4.00 to $5.00 for each $1.00 they invested. After the program ended, 80% of the farmers continued the method on their own.

Reducing nonpoint runoff from other activities

- Soil, oil, and grease are common components of construction site runoff. Lowering runoff of these materials may entail building a settlement pond to trap soil that would otherwise wash away from these sites with rain. Another approach is placing a hay dam or fabric fence around the site to trap runoff. Proper layout of large construction sites to use or modify the land's natural contours can also reduce runoff.
- Soil and soil nutrients eroded from timber-cutting operations and carried into water bodies can be reduced by leaving a strip of uncut trees as a buffer near a stream to absorb runoff. Another means is to build logging roads in a way that minimizes runoff. In some cases, wetlands are built around timber-cutting operations to capture runoff; natural wetlands are also sometimes used.

- Mining operations sometimes produce severe water pollution from runoff that both is acidic and contains heavy metals. Modifying mining sites to allow grass and other plant growth can minimize runoff. A permanent solution is sealing the mine. Runoff from strip (surface) mining sites presents a tremendous challenge. Not only is runoff a major problem, but large land areas often need to be restored as a result.

DRINKING WATER

Drinking water purity is regulated under the Safe Drinking Water Act (SDWA), first passed in 1974 and most recently amended in 1996. This law, which sets health-based standards for microbes and toxic substances that are enforceable by law, ensures uniform health standards nationwide. It also has groundwater protection provisions.

What drinking water pollutants are of concern, why are they a concern, and what are their sources?

About 50% of Americans obtain their drinking water from surface sources, such as lakes, rivers, and streams. The other 50% comes from groundwater. Of the two, surface water is more likely to have significant contamination, but groundwater is an increasing concern. Some drinking water contaminants such as taste, odor, and color, are primarily aesthetic concerns, but others represent threats to human health. In the latter category are infectious microorganisms and nitrate, both of which are considered to present immediate threats to health when present in drinking water at above standard levels. Potential chronic health problems is the primary concern centering around other pollutants, including the metals lead and cadmium and organic pollutants, such as solvents and pesticides (see Tables 7.1. and 7.2). Some are naturally occurring, including the radioactive chemicals radon and radium. Except where spills have occurred, pesticides are not normally found in drinking water at levels high enough to cause acute human health concerns. Again, concern centers around possible chronic effects of low concentrations.

Anthropogenic sources of drinking water contaminants include all the point and nonpoint sources noted in the discussion of wastewater; additional contaminants result from the treatment and distribution of drinking water. In the early years of this century, the major reason that communities treated drinking water was to kill harmful microorganisms. Over the years treatment evolved to include clarifying the water and removing components that contribute to unpleasant taste or odor or to water hardness. (*Hard water* contains dissolved mineral salts, such as calcium and magnesium carbonates, that prevent soap from

TABLE 7.1 National primary drinking water standards (examples)

CONTAMINANT	MCLG (mg/L)	MCL (mg/L)	SOURCES OF CONTAMINANT
VOCs			
Benzene	0	0.005	Some foods, gas, drugs, pesticides, paint
Carbon tetrachloride	0	0.005	Solvents and their degradation products
p-Dichlorobenzene	0.075	0.075	Room and water deodorants, mothballs
Inorganic chemicals			
Barium	2	2	Natural deposits, pigments, epoxy sealants
Cadmium	0.005	0.005	Galvanized pipes, natural deposits, batteries
Nickel	0.1	0.1	Metal alloys, electroplating, batteries
Nitrate	1	1	Animal waste, fertilizer, natural deposits
Organic chemicals			
Atrazine	0.003	0.003	Runoff from use as herbicide
Chlordane	0	0.002	Leaching from soil treatment for termites
Chlorobenzene	0.1	0.1	Waste solvent from metal degreasing
PCBs	0	0.0005	Coolant oil from electrical transformer
Benzo[a]pyrene	0	0.0002	Coal tar coatings, burning organic matter

sudsing or that cause deposits.) Fluoride is added to drinking water in about half of United States communities to lower the incidence of dental caries among water consumers, especially children. Basic drinking water treatment steps are shown in Figure 7.3, but additional steps may be necessary if pollutants are found at concentrations above the health-based standard.

Primary standards. For a pollutant that poses health concerns, EPA must set a primary standard, a process that involves two steps. A *maximum contaminant level goal* (MCLG), the level expected to cause no adverse health effect even over a lifetime of exposure, is first determined. Because it is not always possible to achieve the MCLG, EPA also sets an achievable *maximum contaminant level* (MCL), which is set as close to the MCLG as possible. The MCL is an enforceable standard, which has been set for 83 pollutants (see Table 7.1 for examples). If any of these is detected during drinking water monitoring at greater than its MCL, it must be reduced by means of best available technol-

TABLE 7.2 National secondary drinking water standards (examples)

CONTAMINANT	SUGGESTED LEVELS	CONTAMINANT EFFECTS
Aluminum	0.05-0.2 mg/L	Discoloration of water
Chloride	250 mg/L	Salty taste and corrosion of pipes
Color	15 Color units	Visible tint
Manganese	0.05 mg/L	Taste or staining of laundry
Sulfate	250 mg/L	Salty taste, laxative effects
Total dissolved solids	500 mg/L	Taste or ability to damage plumbing and limit effectiveness of detergents
Odor	3 Threshold odor	"Rotten egg," musty, or chemical smell
pH	6.5 8.5	Low pH: bitter metallic taste, corrosion High pH: slippery feel, soda taste, deposits

ogy. Absorption by activated carbon is a common means used to reduce the level of a number of pollutants. Because some regulated substances – arsenic is an example – also exist naturally, EPA standards take natural as well as anthropogenic sources into account. Primary standards are largely set to respond to potential chronic health effects. However, two contaminants pose an immediate threat to health if primary standards are exceeded. These are nitrate and bacteria and other microorganisms.

Nitrate and microorganisms. Nitrate is a nutrient found in fertilizer. It is most often detected at levels of concern in groundwater in agricultural areas where it has infiltrated wells as a result of fertilizer use. Infants are most vulnerable. Nitrate is converted to nitrite in the infant stomach more readily than in the stomach of an adult or older child. After the nitrite is absorbed into the bloodstream, it reacts with the iron in the infant's hemoglobin to produce methemoglobin, a form of hemoglobin unable to carry oxygen. The result may be "blue baby" syndrome. Excessive ingestion of nitrate has resulted in infant deaths in the Midwest.

The other contaminant that may pose an immediate threat to health is microorganisms. Destroying pathogenic microorganisms remains the primary reason to treat drinking water. Bacteria are naturally present in water; however, fecal coliform bacteria may indicate the presence of human or animal wastes. Coliform do not necessarily cause disease themselves, but rather serve

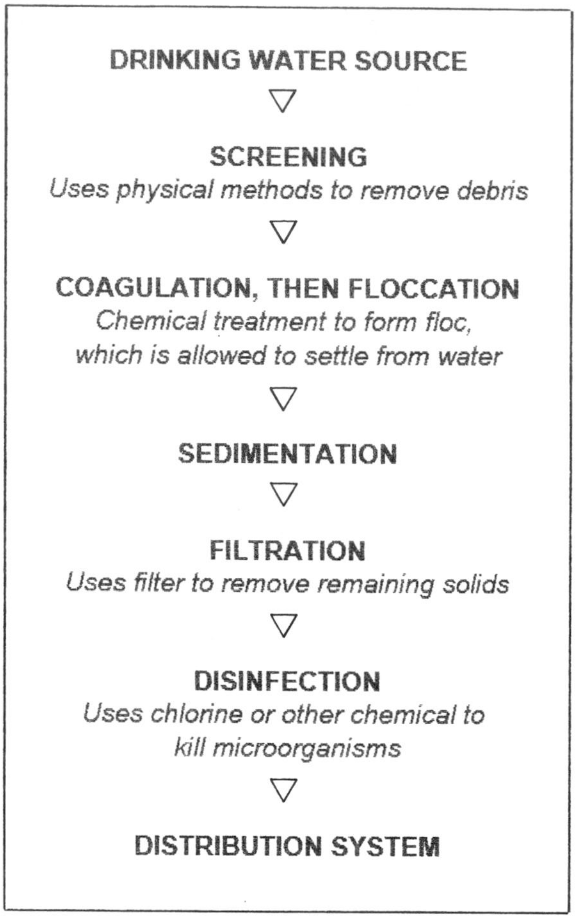

DRINKING WATER SOURCE

SCREENING
Uses physical methods to remove debris

COAGULATION, THEN FLOCCATION
*Chemical treatment to form floc,
which is allowed to settle from water*

SEDIMENTATION

FILTRATION
Uses filter to remove remaining solids

DISINFECTION
*Uses chlorine or other chemical to
kill microorganisms*

DISTRIBUTION SYSTEM

FIGURE 7.3. Simplified flowchart of drinking water treatment. *Source*: Adapted from *Preventing Waterborne Disease*, U.S. EPA, Office of Research and Development, EPA/640/K-93/001.

as an indicator that pathogens may be present. Coliform are used as an indicator because simple inexpensive analytical methods are available to detect them; if they are found in drinking water, it is important to test samples for pathogenic microorganisms. Examples of pathogens are *Legionella*, bacteria that can cause Legionnaire's disease (a respiratory disease); enteric viruses that can cause gastrointestinal distress; and *Giardia lamblia*, a protozoan, which causes intestinal illness. Any of these diseases can be serious or, sometimes, deadly.

Sources of pathogens in drinking water are incompletely treated sewage from sewage treatment plants, septic systems, storm water runoff, and domestic animal manure. Pathogenic microbes may also come from wild animals, including fish and water birds, or from wild animal feces. In early America, contaminated drinking water frequently caused epidemics of cholera, typhoid

fever, amoebic dysentery, and other diseases. As communities started to disinfect their drinking water in the early twentieth century, these diseases largely
ended. Much of the increased life expectancy experienced by Americans in
this century has been attributed to the elimination of waterborne diseases;
nonetheless, pathogenic contamination still occurs. A 1985 United States
CDC report estimated that 940,000 Americans become ill each year from infectious organisms in drinking water, and about 900 die. Infants, elderly people, and those whose immune systems are compromised by other illnesses are
most at risk. An especially serious incident occurred in 1993 when several
hundred thousand people in Milwaukee became ill after drinking water contaminated with the parasitic protozoan, *Cryptosporidium*. More than 100 people who were already ill, and thus had weakened immune systems, died.

Although the United States has problems, most Americans feel comfortable
drinking tap water. The situation is much more serious in many places in the
world where waterborne diseases remain a critical problem. At least 1 billion
people lack access to even minimally safe drinking water. There are locales
where whole villages, especially the children, suffer debilitation from chronic
waterborne infections. Up to 80% of infectious diseases in third world countries are attributed to unsafe drinking water, and, according to the World
Health Organization, more than 35% of all deaths in these countries are directly related to contaminated drinking water. Worldwide, waterborne diseases are estimated to cause hundreds of times more illness than chemical
contamination. For example, the bacterium *Vibrio cholera* in the drinking water of Latin American countries started a cholera epidemic in 1991. Hundreds
of thousands were sickened, about 10,000 died, and cholera is now once more
disease found in Latin America.

Secondary standards. Secondary standards are designed to protect public
welfare. They are guidelines for substances that affect the water's aesthetic
qualities, such as taste, odor, and color, but do not pose a health risk. As guidelines, they are unenforceable, but EPA recommends them as reasonable goals.
Individual states may choose to treat EPA's guidelines as enforceable standards. See Table 7.2 for examples of secondary standards.

Disinfection by-products. In recent years, concern has arisen about disinfection by-products (DBPs) formed during chlorine disinfection of water. Organic material is present in water to varying extents and reacts with the chlorine disinfectant to produce DBPs. Chloroform is the DBP produced in the
largest amount. At high doses, chloroform is an animal carcinogen, and epidemiologic studies suggest that drinking chlorinated water for many years increases the risk of bladder and rectal cancer. This raises a dilemma because
disinfection is critical to killing harmful microorganisms and levels of disin-

fectant high enough to accomplish this must be maintained, but the higher the concentration of chlorine, the greater the amount of DBPs produced. Some argue that chloroform concentrations and those of other DBPs are too low to raise concern, especially when balanced against the development of infectious diseases.

EPA has established an MCL for total trihalomethanes (TTHMs), a family of DBPs found in chlorinated water. There are several possible means to reduce DBPs:

1. Remove DBPs from drinking water after they are formed; this is end-of-pipe control and is costly.
2. Remove as much organic material from the water as possible before it is chlorinated, thereby allowing formation of fewer DPBs, and use as little chlorine as is consistent with maintaining disinfection.
3. Use an alternative disinfectant that produces smaller amounts of DBPs than chlorine. The alternate disinfectants that follow do produce smaller amounts of DBPs; however, some of these have not been characterized and may also pose health problems.

   • Chlorine dioxide: This chemical effectively disinfects water while producing much smaller quantities of DBPs; it is used by an increasing number of communities.
   • Chloramine: This effective disinfectant, used by some communities, also produces fewer DPBs.
   • Ozone: Ozone disinfection works even better than does chlorine, but it breaks down to oxygen, which cannot maintain disinfection, so a small amount of chlorine must be added before piping the water to users. This produces DBPs, but in much smaller amounts. Many consider ozone an ideal disinfectant. However, one of its DBPs, bromate, is a strong animal carcinogen. Ozone disinfection is also more expensive and operators need to be more highly trained.
   • UV light: This provides a nonchemical way to disinfect water but is very expensive and is used primarily to treat small volumes of water. As with ozone, small amounts of a chemical disinfectant must be added before water is piped to users.

Water disinfection and purification provide another example of a solution that is not so much environmentally friendly as less unfriendly. The best means to reduce DBPs and other water pollutants is, as always, P^2. If fewer infectious microorganisms enter the water, less disinfection is needed. If fewer potentially toxic chemicals enter the water, fewer have to be removed. EPA has established a watershed protection approach to maintain the purity of the water supply and is working with the states to develop specific programs. The ma-

BOX 7.6

When is fluoride a pollutant? About half of all Americans drink water to which fluoride has been added to prevent dental decay. Fluoridation has been practiced for 50 years but remains controversial because some believe it is harmful to health. A 1994 CDC report stated that fluoride, at typical exposure levels, does not cause bone disease, cancer, infertility, birth defects, or the other ills that have been attributed to it. EPA's MCL for fluoride in drinking water is 4 mg/L (4 ppm). However, the MCL is higher than the 0.7–1.2 mg/L considered optimal to prevent caries, and people also ingest fluoride from food and beverages processed with fluoridated water. Fluoride is sometimes found in tooth paste, which children may swallow. In few people who ingest more than 2 mg/L stained or pitted teeth have developed. In a few cases, in people who drank water that naturally contained more than 8 mg/L of fluoride for many years a crippling skeletal disease developed. Thus, while supporting water fluoridation, the CDC noted the need for care in its use and for more research.

jor water pollution problem is runoff from nonpoint sources. Thus, the P^2 approaches needed are the same ones used to reduce the various sources of rural and urban runoff. However, as population and development pressures grow, maintaining water purity by whatever means becomes more difficult.

Maintaining drinking water standards. Almost all states have *primacy* in enforcing SDWA regulations, meaning that EPA has given to the states the right and responsibility to enforce federal drinking water regulations. Unfortunately, the United States General Accounting Office reported in 1993 that many states are not fulfilling requirements. For example, although states are required routinely to evaluate the performance of water supply systems, an increasing number fail to do so because of limited resources. This situation is worsened by the fact that water supply systems are aging and often lack the modern technology needed for optimal performance. Nearly 90% of United States public water systems are small, serving 25 to 3,300 customers. In earlier years, small systems only had to meet standards for water clarity and bacteria and nitrate levels; however, since 1986 they, like their larger counterparts, must frequently test for more than 80 pollutants unless they can prove that particular substances are not found in their watershed. For small systems, monitoring costs are a burden, and, if treatment is needed to remove a pollutant, costs can be prohibitive. Since federal money is not available to help meet regulatory requirements, communities have begun to resist EPA mandates. In response, the Clinton administration, in late 1993, proposed to make the Safe Drinking Water Act more flexible for small communities. Under this plan, communities would be asked to develop and to carry out a P^2 strategy to protect drinking water supplies. Those doing a conscientious job would be exempted from some monitoring and treatment requirements.

NEAR-COASTAL WATERS AND ESTUARIES

Contamination of all water bodies is a concern and each raises special problems. Near-coastal waters and estuaries pose special concerns. About half of all Americans live within 50 miles of a marine or Great Lakes coastline. Many near-coastal waters and estuaries are poorly protected from the activities of this large and growing population. About 85% of this country's 10 billion daily gallons of wastewater flows into estuaries and bays, and, not surprisingly, sewage pollution is common, especially near large cities. Shellfish beds are often closed because of contamination with fecal or toxic microorganisms; metals and toxic organic pollutants can also be serious contaminants. As is the case with rivers and lakes, a significant portion of near-coastal and estuary pollution arises from rainwater and snow-melt runoff. Wildlife and bird populations have decreased in some areas along coastlines. Some of this decrease is due to habitat destruction, but pollution plays a role. EPA is working with Great Lakes and coastal states and cities to reduce pollution enough to allow the lifting of fishing bans and advisories and to allow shellfish to be safely harvested from the one-third of beds now closed. Dumping of sewage sludge and industrial waste into oceans has been banned, and in the future large coastal cities will be required to have storm water discharge permits. Industry and municipalities will also be required to meet stricter CSO regulations. Oil spills are a special problem in coastal areas. The Oil Pollution Act, passed in 1990, requires double hulls for new or upgraded oil-carrying vessels and imposes standards for them. At least some oil companies are making safe transport a major priority, but oil spills continue as a result of human error, bad weather, and crowded harbors. True P^2 would mean using less oil and more renewable energy.

In some coastal areas, pollution may have decreased in recent years. The National Oceanic and Atmospheric Administration (NOAA) reported that chemical contamination of about 300 American coastal sites was lower in 1991 than in 1984 when this agency first monitored shellfish, bottom-feeding fish, and sediments for metals, PCBs, PAHs, and the pesticides DDT and chlordane. Dramatic drops were seen in levels of DDT, PCBs, and lead, all of which were banned in the 1970s. Not surprisingly, the most heavily contaminated waters were found adjacent to large cities, such as Boston, New York, and Los

Angeles. Even these waters did not contain contaminant levels considered toxic to individual marine animals; however, some fear the contamination is sufficient to affect reproduction or to weaken the immune system of marine animals. Still, the NOAA results are heartening because they show that environmental regulations have had positive results. Also, as environmental levels of persistent chemicals decrease, animal body burdens will slowly decline and, with them, the fear of reproductive or immune system toxicity.

The NOAA researchers did not test coastal areas that were already known to be highly contaminated, but sampled coastal conditions more broadly. However, the United States FDA recently studied seafood contamination at a *hot spot* (that is, a heavily contaminated area) about an hour's sailing time from Boston. PCBs, many other chemicals, and radioactive materials had been dumped there for many years before 1976. In 1992, FDA collected lobsters and other bottom dwellers at this spot along with many fish varieties, including cod, flounder, and ocean trout and tested them for PCBs, PAHs, heavy metals, and pesticide residues. Pesticide residues were detected in only a few seafood samples. About half the samples had no detectable PCBs, and the rest had either trace amounts or levels within acceptable FDA limits. Cadmium, mercury, lead, and arsenic levels were all safely within FDA limits. A few seafood samples had radioactive contamination, but the type of radioactive components present indicated that they probably resulted from 1960s atmospheric testing rather than dumping. An FDA spokesperson noted, "This snapshot concludes that the overall residues are low and the seafood from Massachusetts Bay is safe to eat." Both the NOAA and FDA studies show that contamination of water bodies and seafood decreases when polluting activities are eliminated or decreased. Although these reports are encouraging, high concentrations of PCBs and other chemicals still survive in the sediments of many hot spots around the Great Lakes, the marine coasts, and some rivers. However, despite the environmental persistence of PCBs, they do slowly degrade. A larger challenge will be the continued protection of coastal areas as human populations and development activities continue to increase.

An example of problems faced by near-coastal waters. One hundred fifty rivers and streams from six states and the District of Columbia compose the Chesapeake Bay watershed. Like many other bays, the Chesapeake was once a flourishing ecosystem and a major source of fish and other seafood. However, over decades of increasing population and industrial activity, bay water quality deteriorated. Fish and oyster populations dropped as much as 80%. Underwater grasses disappeared. The major pollutants found in the bay are nutrients, toxic substances, and microorganisms. Nutrient sources are wastewater treatment plants and runoff from farms, homes, and septic systems; toxic pollutants arise from industrial point sources and urban runoff. Sources of microorganisms are improperly treated sewage and animal waste. Not all runoff is from coastal land; rivers, for example, carry large amounts of nutrients and other pollutants into the bay.

Although the most severely polluted areas are around urban centers, the whole bay has been adversely impacted to some extent. With EPA support, the Chesapeake Bay Commission is working with the states involved to develop better pollution control and, most importantly, P^2 strategies to protect the estuary and adjacent coastal waters. Reducing nonpoint sources is the chief challenge.

- One major goal is to reduce runoff of the nutrients nitrogen and phosphorus entering the bay by 40% by the year 2000. One method being used to achieve this goal is to train individuals to write nutrient management programs for individual farms that will allow farmers to apply fertilizer, ma-

---

BOX 7.7

Pollution is only part of the problem. This book focuses on pollution, but recall EPA's Science Advisory Board's analysis of environmental priorities and see Table 4.3. Pollution issues figured prominently, but habitat alteration and destruction and species extinction and loss of biodiversity were also major priorities. The authors of a 1995 *Science* article on American freshwater ecosystems commented that the intent of the CWA and SDWA was good, but that the emphasis on pollution shifted attention away from equally harmful and pervasive forms of environmental degradation such as habitat destruction, and invasions by exotic species. Except in Alaska, water flow in virtually every freshwater body in the United States has been modified by dams and water diversions. These actions have greatly modified ecosystems. For example, rivers are much less able to serve as migratory pathways for fish and other animal species; nor, with their interrupted flow, can rivers maintain downstream ecosystems. Meanwhile, water consumption in the United States has more than doubled since 1940 and is likely to double again in the next 20 years, intensifying the already great pressure on water bodies. The CWA and SDWA do not deal with these and other very important issues. Although it is critical to protect water bodies from pollution, we also need to maintain and restore important ecosystems. More generally, we need to analyze our use of water critically with a willingness to make the changes needed.

nure, and sludge to their land in a manner that will prevent the nutrients from being carried away in runoff.

- To reduce pesticide runoff, voluntary programs are being developed. One has the goal of enrolling 75% of farmers and other pesticide users into integrated pest management programs by the year 2000.
- To reduce pathogenic microorganism contamination, wastewater treatment plants are being upgraded and plans made to control millions of home septic systems.
- To reduce toxic chemical input, a program is working on strategies specifically designed to reduce input into the bay of 14 high-priority toxic substances.
- Another program is working to provide forest buffers on river banks to absorb the contaminants in nonpoint runoff.

Some encouraging early results have been seen in Chesapeake Bay. Underwater grass beds have returned and a record number of striped bass are now in the bay. There has also been a 52% reduction in industrial releases of toxic substances.

GROUNDWATER

About 50% of Americans, especially rural Americans, rely on groundwater for drinking. Groundwater is found beneath the earth's surface in aquifers (geologic formations), where water has collected in quantities sufficient to supply above-ground springs and wells. Although sometimes groundwater flows in a channel, an aquifer is not ordinarily like a surface stream; rather, it is a formation composed of permeable rock, gravel, or sand that is saturated with water. Groundwater in these formations can flow, but it ordinarily moves more slowly than surface water. Anything that contaminates surface water can potentially contaminate groundwater. Surface pollutants, dissolved in water, percolate down through the soil and, depending upon soil and pollutant characteristics and the distance to groundwater, they may reach groundwater. Sources of contamination include agricultural and urban runoff, chemical spills, and landfill leachate. Pollutants of special concern are sewage from improperly installed or maintained septic systems and petrochemicals from leaking underground storage tanks. Much American groundwater has detectable levels of pesticides and other contaminants. Detectable levels do not necessarily indicate a problem, but they do indicate that ongoing monitoring and efforts to prevent further pollution are important to assure that levels do not increase.

EPA is working with the states to develop P^2 strategies to protect the watershed that feeds an aquifer or wellhead. The word *wellhead* defines a more limited area than a watershed; it is the immediate area around a public water sup-ply intake. An example of how P^2 can be used to maintain groundwater purity in a wellhead area is to specify which pesticides can be used and the manner of application that will result in the least runoff. For example, a pesticide should have little tendency to migrate into groundwater. Another P^2 approach in the wellhead area is to specify how land can be used: Landfills may no longer be sited over groundwater; or, siting a gasoline station over groundwa-ter that feeds into a wellhead can be prohibited. A more controversial future possibility is to prohibit farmers from grazing livestock on land over vulnera-ble groundwater.

Cleaning up groundwater

Once polluted, groundwater is extremely costly to clean up although there are many locations in the United States where cleanup is needed. Often, the greatest concern about a hazardous waste site is that it may contaminate groundwater used for drinking. Cleanup is often not feasible with today's technology, although a pump-and-treat process has been commonly used. In this procedure, the water is pumped to the surface, treated to remove pollu-tants, and then returned to its source. Once begun, pump-and-treat may con-tinue for many years with the goal of attaining drinking water quality. It has sometimes been used simply for lack of better alternatives. An NAS panel, af-ter an extensive study of pump-and-treat, concluded that some sites would not reach drinking water quality even if treatment continued for 1,000 years. In a 1994 report, the panel advised against routine use of pump-and-treat and in-stead recommended that the contaminated water be contained in place until some future time when effective treatment technologies become available. Containment in place involves building an underground structure to prevent the contaminated water from migrating off-site.

In 1995, reports of an effective, quite simple, and cheaper alternative to pump-and-treat emerged. Most of us have seen the reddish iron rust formed when iron is exposed to the oxygen in air. In an analogous way, the new tech-nology allows iron to react, not with oxygen, but with groundwater pollutants. Tons of iron filings mixed with sand are installed underground in such a way that the contaminated groundwater must flow through it; as it does, pollutants react with the iron. Trichloroethylene is a common groundwater contaminant that is effectively treated in this manner; it reacts with iron and is degraded into benign products. Once past the barrier, the remediated groundwater con-tinues on its course. This technology is not a cure-all and cannot destroy some difficult pollutants, for example, PCBs. However, it is one promising alternative to pump-and-treat.

WETLANDS

Both coastal and inland wetlands (swamps, bogs, marshes) perform valuable pollution control functions. They absorb microorganisms, sediment, nutrients, and other pollutants that run off into them, and wetland plants and microorganisms often degrade these pollutants. These properties make wetlands valuable protective buffers between land and coastal areas. Indeed, constructed wetlands are sometimes built around forestry operations to trap runoff, and a number of small communities use constructed wetlands as an alternative to traditional wastewater treatment. In addition to their valuable pollution control role, wetlands offer habitat for thousands of species of fish, other aquatic organisms, and water birds and animals. They also serve as flood control areas. Wetlands that are wet only part of the year perform equally important functions (see T. Williams's article, "What Good Is a Wetland?")

DISCUSSION QUESTIONS

1. The population of southeastern Michigan is projected to grow 6% in the next 20 years, but the amount of developed land will increase 40% if present trends are left unchecked. Nationwide, the average number of persons per household has declined from 3.6 to 2.7 over the past 40 years leading to a significant increase in the total number of households. Other development activities that have been increasing include an increasing number of vacation homes and an increasing amount of shopping space per person (usually in the form of more malls).
   (a) How do these activities lead to increased water pollution?
   (b) How do they lead to increased air pollution?
   (c) How do they lead to increased land pollution?
2. Environmental degradation is a major problem in some cities. However, a city like New York can produce fewer pollutants per person than a suburb. How?
3. (a) Why is pollution of estuaries and other near-coastal waters such a major concern?
   (b) From your perspective, how can they be better protected?
   (c) Address the same questions in relation to groundwater pollution.
4. Examine Figure 7.4.
   (a) What actions can you take within your own home and yard to reduce the amount of water that you use?
   (b) What actions can you take to reduce the amount of pollution in the wastewater that your home produces?

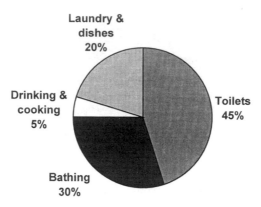

FIGURE 7.4. Water use in the home. Data from
Pipeline, 6(4), Fall 1995.

TABLE 7.3   Growth in the world's population

BILLIONS OF PEOPLE	YEARS TO ADD 1 BILLION
1	1,000,000
2	105
3	30
4	15
5	12
6	11

Source: World Watch, 8(6), Nov./Dec. 1995, p. 18.

5. Examine the information in Table 7.3, which shows that the world's
   population is now close to 6 billion people.
   (a) At the current rate of growth, about how many years will it take
       the world's population to reach 7 billion?
   (b) How optimistic are you that we can meet the demands made on
       the environment by increasing population growth? Explain.

FURTHER READING

Adler, T. 1996. The Expiration of Respiration: Oxygen the Missing Ingredient in Many Bodies of Water. *Science News*, 149(6), 88–9, Feb. 10.

American Chemical Society. 1989. *Ground Water, an Information Pamphlet*. Washington, D. C.: ACS.

Anderson, H. A. 1990. Devising Advisories: Are Fish Safe to Eat? *Health & Environment Digest*, 4(10), 1–3, Dec.

Fisher, B. E. 1994. Downstream in America. *Environmental Health Perspectives*, 102 (9), 740–45, Sept.

Hinrichsen, D. 1996. Coasts in Crisis. *Issues in Science & Technology*, XII(4), 39–47, Summer.

Hoffman, D. A. 1991. Fluoride: Risks and Benefits. *Health & Environment Digest*, 5(3), 1–4, April.

Meadows, R. 1995. Livestock Legacy. *Environmental Health Perspectives*, 103(12), 1096–1100, Dec.

Smith, V. 1994. Disaster in Milwaukee: Complacency Was the Root Cause. *EPA Journal*, 20(1–2), 16–18, Summer.

Steiner, R. 1994. Oil-Stained Legacy. *National Wildlife*, 32(5), 37, Aug./Sept.

U.S. EPA Office of Water. 1994. *Is Your Drinking Water Safe?* EPA 810 F 04 002, May.

U.S. EPA Office of Water. 1994. *Office of Wastewater Management Primer*. EPA 830-K-94–001, Nov.

Wilcox, K. 1996. Alternative Toilets: To Flush, or Not to Flush. *Small Flows*, 10(1), 6–7, Winter, 1996.

Williams, T. 1996. What Good Is a Wetland? *Audubon*, 98(6), 42–53, Nov.-Dec. 1996.

8

SOLID WASTE

The 1970s and 1980s saw major changes in the way Americans handle their wastes. An increasing environmental awareness led to the passage of the 1976 Resource Conservation and Recovery Act (RCRA, pronounced "rick-rah"), a law governing management and land disposal of solid waste. RCRA is a complex law, and, having passed it, Congress believed that this country's waste management practices were well covered. However, shortly afterward, there were revelations of abandoned hazardous waste sites around the country. Congress addressed these in 1980 through the Comprehensive Environmental Response, Compensation, and Liability Act (CERCLA, commonly called Superfund). CERCLA governs the cleanup of abandoned sites containing hazardous wastes that have escaped into the environment or have potential for doing so. Over the years, both RCRA and CERCLA have been significantly amended. This chapter focuses on municipal solid waste (MSW) and hazardous waste, but it is important to recognize that these are a small proportion of total wastes. MSW may be as little as 2% of the solid waste generated in the United States and about 3% of hazardous wastes. Other solid wastes that RCRA regulates are construction and demolition debris, municipal sludge, combustion ash, mining and drilling debris, agricultural wastes, and some industrial process wastes that can be disposed of in MSW landfills.

MUNICIPAL SOLID WASTE

Americans consume about 25% of the world's resources. Not surprisingly, we also generate more waste per person than any other country; on average, each of us produces about 4.4 pounds of trash a day. After recovery, recycling, and composting, this figure is 3.4 pounds a day. The total amount of trash discarded by Americans in 1995, about 210 million tons, was more than twice that of 1965 (review Figure 2.1 and see Table 8.1). The cost of managing MSW was about $30 billion in 1991, projected to increase to $75 billion by the year

TABLE 8.1 How much waste do we throw away?

POUNDS/PERSON/YEAR	
United States	1,580
Japan	900
Western Europe (average)	770
Netherlands	1,100
Portugal	570

Source: EPA Waste Characterization Report, Franklin Associates, 1992.

2000. Even at $30 billion, community spending on trash is second or third only to spending on education and police protection. Increased costs are due to the greatly increased care with which we manage waste. Not everyone believes that we are producing inordinate amounts of waste. William Rathje, a

---

BOX 8.1

Space ecology. Manned space flights now store their wastes to take back to earth, but this practice would not be practical for long-term missions. In 1978, NASA started researching a self-sustained system for living in space, the Controlled Ecological Life Support System. In this system, there are no wastes, only resources. Green plants would be grown and harvested as food for the astronauts. Plants would absorb carbon dioxide that humans respire and emit the oxygen that humans need. Plant transpiration would provide pure water. As biological resources, plant and human wastes would be recycled with the assistance of microorganisms and of technologies that can convert the wastes to forms plants can use. A computer would integrate this complex system, which would first be tested on earth. Knowledge gained in developing this system is expected to have applications not only in space, but on an earth that contains more human beings, more wastes, and fewer virgin resources.

garbage archeologist, does not believe that Americans produce more trash than did our ancestors 50 or 100 years ago. He believes there is more waste today only because our population is larger. In a recent book, *Rubbish: The Archeology of Garbage*, Rathje states: "All of America's garbage for the next 1,000 years would fit into a space 120 feet deep and forty-four miles square – a patch of earth representing less than 0.1% of the surface area of the United States."

What are the pollutants of concern in MSW and the sources of MSW?

Figure 8.1 shows the composition of United States discards in 1995. Paper represented the largest proportion of the material in MSW, 33% by weight, followed by yard wastes, plastic, food waste, metal, and glass. The "other" category includes appliances (white goods), furniture, and batteries. Household hazardous waste, leftovers of products like paints, pesticides, and automotive maintenance items, also falls in this category. Although industrial hazardous wastes are carefully regulated and not allowed into the MSW stream, no regulations are imposed on those generated by individual households. Householders may not even think about the hazardous nature of many products they use. A single household throws away only small amounts of hazardous waste, but when they are multiplied by 100 million similar households, significant amounts are generated. Social and economic status affect what we throw away. Upper-income groups throw away a lot of yard waste, such as pesticides, herbicides, and fertilizers. The home-improving middle class disposes of more unused paints, stains, and varnishes. Low-income people discard more used

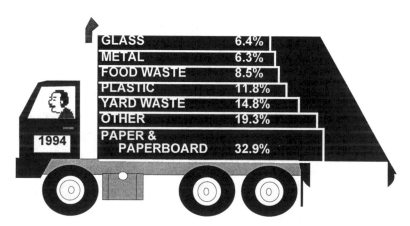

FIGURE 8.1. Composition of wastes thrown away by Americans. *Source: MSW Factbook*, Ver. 3.0, Office of Solid Waste, U.S. EPA, Washington, D.C., 1996.

motor oil and auto maintenance items along with more clothing, often reused clothing bought at used-clothing shops or yard sales. Poorer people cannot afford to buy in bulk and more often buy smaller sizes of cereal, detergent, and canned food. This means they discard more packaging than those who can afford the more efficiently packaged larger size.

Some states treat construction and demolition debris as MSW, but the federal definition does not; this debris equals about 12% of MSW by weight. The definition of MSW also does not include municipal sewage sludge or combustion ash, although these are sometimes landfilled or incinerated along with MSW. Sources of MSW include wastes generated by households; commercial sources, including restaurants, grocery stores, and offices; and institutions, schools, hospitals, and prisons.

Why is MSW of concern?

Even if it had no other undesirable characteristic, the sheer amount of trash generated presents concerns: How do we handle it and where should it go? One household of four people discards an average of 17.6 pounds per day (4 × 4.4); over a year, this amounts to 6,400 pounds. In earlier years when most families lived in rural locations, this amount of waste and more could be disposed of with little concern. But if individual households today had to deal with their own wastes, a crisis would quickly develop. It is no longer a feasible option for most. Garbage is also unpleasant; diapers and other sanitary items have infectious microorganisms and unpleasant odors, as do rotting food and yard waste. The hazardous wastes discarded by householders can injure workers who collect or process them if they catch on fire, give off fumes, or contact workers' skin or eyes. Hazardous wastes have also damaged facilities that handle them. Combustion and landfilling of MSW carry their own risks, which will be discussed later in this chapter.

How is MSW being reduced?

As the name of the Resource Conservation and Recovery Act implies, the intent of RCRA was to stimulate communities to manage waste in an environmentally sound way and to recover materials and energy. Recall the waste management hierarchy described in Chapter 2. Waste reduction (also called source reduction or P²) is at the top of the hierarchy, followed by reuse and recycling (including composting). Treatment to reduce waste volume or toxicity is third on the hierarchy. For MSW, treatment usually means combustion in waste-to-energy plants. At the bottom of the hierarchy is disposal. For MSW, this ordinarily means landfilling. Each step of the hierarchy will be discussed.

BOX 8.2

A paradox. An examination of the products at a grocery or other retail store in the United States would quickly reveal many examples of products with "excess" packaging. The amount of excess would depend on individual definitions. However, reducing the amount of packaging is not always desirable; for example, reducing the amount of packaging used on food seems an obvious plus, but proper packaging reduces food spoilage and thus reduces the amount of food thrown away. Americans dispose of 15% to 20% of food as waste from the time it leaves the farm until it reaches the consumer's table. Compare this to very poor countries where foods lack protective packaging. Between 40% and 70% of the food is thrown away because of spoilage during shipping and storage.

Pollution prevention. Illustrations of P^2 were given in Chapter 2, and this important subject will be emphasized again. Even small P^2 efforts can have far-reaching impact on MSW. Several examples follow:

- Concentrated cleaning products: A more concentrated laundry detergent has the same cleaning power as a smaller weight of detergent in a smaller container. Because the consumer uses less detergent, less of the product's components is discharged to the sewer. Because the same amount of cleaning power is packaged in less space, there is less packaging waste to be discarded. Less packaging also means fewer resources are used to make the packaging; resources are conserved. Because detergent weight per unit cleaning power is lowered, it takes less energy to transport a given amount of cleaning power, resulting in resource conservation. Using less energy in transportation means fewer pollutants are emitted during transport. Another advantage of the concentrated detergent is that less packaging may cost the manufacturer less to buy and transport.
- Light-weighting containers: While still maintaining container strength, manufacturers reduced the weight of plastic soda bottles 25% and the weight of glass bottles 31% between the late 1970s and 1993. This has many of the advantages noted for concentrated detergents.
- Photocopying both sides of paper: This cuts the amount of paper used in half – conservation – and also halves the amount discarded as waste. Other ways to limit paper use are to post a notice on bulletin boards rather than sending individual notices to many people and using electronic mailings.

Reducing the toxicity of MSW is another goal of waste management. Newspaper and magazine printers traditionally used heavy metals: lead ink for print, and heavy metal pigments for colored sections. However, as soy bean and organic inks have been developed, heavy metals have been eliminated. Heavy metals also serve useful functions in plastics and in packaging, but manufacturers of these products are under pressure not to use them. Manufacturers have found relatively benign organic molecules that can serve the same function, with the result that plastics and packaging contribute smaller amounts of heavy metals to MSW. Nonetheless, plastic still represents the second largest source of the heavy metal cadmium in the American waste stream. Finding substitutes for heavy metals in these products required a partial redesign. As such, these are examples of the emerging process of *design for the environment* (DfE). There is a growing recognition that, for P², recycling, and reuse to be truly successful, there must be fundamental changes in product design. The following section discusses this important form of P² in more detail.

Design for the environment: In conventional design, if industry worked to minimize energy input, it did so to save money. Because energy was typically inexpensive, using energy efficiently was not a priority. The pollution produced by energy usage was taken by industry and by society as a given. Unless obvious monetary savings were involved, using raw materials efficiently was likewise not a priority. If the amount of product was obviously increased by minimizing waste, industry worked to minimize waste, but, again, not ordinarily as a priority. Green design, or DfE, is different: From the moment a new product is conceived, it is designed to lessen its environmental impact (see Figure 8.2). DfE considers – at every step in the design of a product – how to use energy and water in the most efficient way, how to use raw materials in the most efficient way (including, for example, recycled materials), and how to prevent the formation of both hazardous and other solid wastes. In summary, those using DfE strive for a product that, during its manufacture (and the manufacture of its component parts), uses fewer material resources, less energy, and less water; produces less pollution; and uses fewer hazardous chemicals.

Removing cadmium from disposable batteries or heavy metals from packaging and newspapers did not involve complete redesign of the products involved. Xerox Corporation provides a more comprehensive example of DfE. Xerox had been taking back its copying machines from rental users since the 1960s and reusing them; this practice was significant, especially as compared to that of other companies of the period. However, Xerox did not design its products in a manner that allowed parts and materials to be easily recovered at the end of product life or that made it easy to remanufacture those parts into new products. In 1990, it changed its approach: From the moment a new product idea is conceived, DfE is used. Criteria were added to a product's design that had the intent of producing a longer-lived product that was easier to

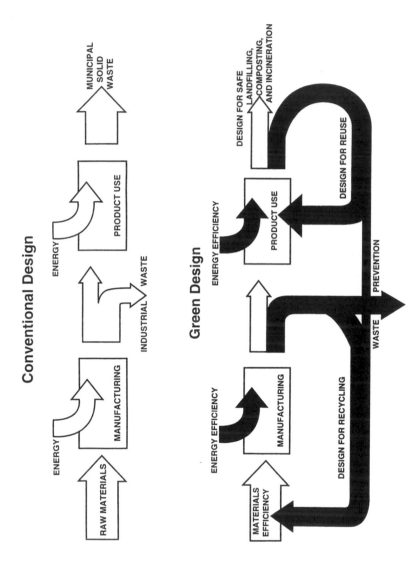

FIGURE 8.2. How product design affects material and energy flow. *Source:* Adapted from U.S. Congress Office of Technology Assessment, *Green Products by Design: Choices for a Cleaner Environment,* Washington, D.C.: U.S. Government Printing Office, 1992.

disassemble at the end of its life, allowing it to be remanufactured into a new copier or manufacturers to recover recyclable parts. Also built into product design is *remanufacturing convertibility*; that is, product components are designed in a way that allows them to be put to a different use. For example, a copier part may find a second life in an electronic printer. Proctor & Gamble, as another example, found, upon analyzing energy consumption over the lifetime of its laundry detergents, that the most energy-intensive step is the heating of washing machine water. Consequently, they began to design detergents that work effectively in cold water or when less water is used.

Sometimes DfE takes totally new approaches. One manufacturer is developing a dishwasher that uses ultrasound in its cleaning process. This washer needs less water, less electricity, and a smaller amount of a less caustic detergent. It also runs more quietly. Although its price is higher than average, this cost will be recovered by energy, water, and detergent savings. DfE is becoming a necessity in Europe, where manufacturers of electronic products will be required to recover and recycle their products. The goal is to divert electronic scrap from the waste stream; this will require manufacturers to redesign refrigerators, computers, stereo equipment, and other electronic products. In the short term, doing so is a tremendous problem for manufacturers because products already on the market were not designed for reuse and recycling. Sometimes even the material composition of products is not known; nor are the purity of potentially recyclable materials or the hazardous constituents that may be present. Plastics present special recycling problems.

Life cycle assessment: DfE can be a very difficult process, but at least it is working with only one or a few related products at a time. It also has very specific goals in mind. However, when one begins to think comparatively about what makes one product environmentally preferable to another, the situation is even more complicated. *Life cycle assessment* (LCA) is a tool being developed to carry out the task of understanding the environmental impact of products. As the name implies, LCA examines a product from cradle to grave: What raw materials go into making the product? How much energy and how much water are needed to acquire the needed raw materials? What air and water emissions go into acquiring raw materials, and how much waste is produced? Similarly, when the product is manufactured, how much energy and water is needed, and what wastes and emissions are produced? As it is used? When it is recycled or disposed? Comparison of products by LCA seldom can definitely answer the question as to which is better. One product may use recycled materials but produce more pollution during manufacture than a competing product. One may need less energy in its manufacture but require more toxic chemicals. Many other examples could be given.

Early life cycle assessments examined relatively simple products and asked questions like, Are paper bags better than plastic bags? and, Are disposable

paper cups better than plastic cups? The conclusion was that plastic was probably no worse than paper. However, some believed that the proper question
should have been, Which is better, a disposable cup or a reusable one? One is
tempted to respond confidently that the reusable cup is preferable; however,
an assessment of this question concluded that a glass cup must be used hundreds of times to be preferable to a disposable cup. The glass cup takes more
resources and energy to produce, generates more pollution during its manufacture and transport to market, and uses additional energy and water each
time it is washed as well as producing pollution during washing. Furthermore,
any one householder usually has many glass cups. If the householder owned
only a few cups that were reused a great many times, then glass cups would
likely be environmentally preferable. Perhaps the real problem is that the
average American uses many glass cups and many disposable cups at the same
time.

Cloth versus disposable diapers present a similar dilemma. Each time cloth
diapers are laundered, water is used, energy is needed to heat the water, wastewater is generated, and, unless they are hung up on a clothesline, energy is
used to dry the diapers. If a diaper service is used, the energy consumption
and pollution produced by transportation to and from the laundry must be included. On the other hand, disposable diapers create more waste than do
cloth diapers. This waste is not easy to recycle, although when it is burned in a
waste-to-energy combustor, some energy is recovered or, if the plastic covering
is removed, the fiber portion of the diaper can be composted. It can be very
difficult to compare even simple products and becomes progressively more
difficult with more complicated products. However, LCA is a technique in the
early stages of its own life cycle and holds the promise of better future product
comparisons.

Some European countries and Canada have begun to use green labels to
denote products that they believe to be environmentally preferable. The
United States EPA is also working to define characteristics of preferable products. Once it decides what those characteristics are, government agencies will
preferentially buy the products identified. It is often difficult for individuals to
make product comparisons. However, one obvious P^2 step – practicing conservation – is straightforward. One small example: Consider that Americans, 5%
of the world's population, use about half of all the detergent produced in the
world. Do we need this much detergent to keep clean? Is it possible to use a little less shampoo or hand soap, or to use a washing machine only with a full
load? Or to purchase a front-loading washing machine, which uses less water
and less detergent (and thus less energy to heat the water) and produces less
wastewater?

Individuals can practice P^2 by looking for and purchasing durable consumer goods or by repairing a television set or appliance rather than discard-

ing it. A refrigerator is a major user of energy in the home. Purchasing a refrigerator that uses less energy than others is P². The way a refrigerator is operated also affects the amount of energy it uses. Purchasing a car with good gas mileage and maintaining it well are other important ways to reduce both energy use and pollutants emitted by the vehicle. Smaller, but in the aggregate, important steps are buying concentrated products when they are available or buying bulk quantities if you are able to use up the product. In these cases, there is less packaging per unit of product. Taking fabric bags to go shopping rather than accepting new bags at the store saves resources and prevents the pollution used to manufacture new products. Even paper bags can be reused as long as they remain intact. There are hundreds of ways to reduce energy use and water use and to produce less household waste. Multiplied by hundreds of millions, the net impact is indeed significant.

DISCUSSION QUESTIONS

1. Think about a cotton shirt. Starting with growing the cotton and ending with transporting the shirt to a retail store, consider the steps involved.
   (a) What is the environmental impact of producing this shirt at each step? Include the impact of wearing and disposing of the shirt.
   (b) Repeat this exercise for a synthetic fabric shirt, starting with recovering the oil needed to make it.
   (c) Consider how you would compare the environmental impacts of the cotton and synthetic shirts. What difficulties arise in the comparison?
2. Examine Figure 8.3. Remember the difference between organic (biodegradable) materials and inorganic (nonbiodegradable) materials.
   (a) Which of the materials shown will biodegrade?
   (b) Which will – at least at first – crumble into smaller pieces, rather than truly degrading?
   (c) If the biodegradable materials were buried in a landfill, what would you expect to happen to biodegradable materials within the time frame shown? Explain.

Reuse. Remember from Chapter 2 that reusing a product is typically superior to recycling it when this option is available because fewer resources are used and less pollution is produced. It is for these reasons that individuals employed in a DfE project attempt to design reusable components in a product that are easily removed when the product is discarded. Consider the difference between reusing a metal-containing component of a product and recycling the metal. To recycle the metal, the component must be melted down

Bottle	1,000,000 years
Plastic six-pack holder	450 years
Aluminum can	200-400 years
Tin can	80-100 years
Rubber sole	50-80 years
Leather belt	40-50 years
Nylon fabric	30-40 years
Hard plastic container	20-30 years
Disposable diaper	10-20 years
Wooden stick	10-15 years
Cigarette butt	2-5 years
Hemp rope	1-14 months
Woolen cap	12 months
Cotton rag	1-5 months
Banana/orange peel	3-5 weeks
Paper	2-4 weeks

FIGURE 8.3. The lasting litter chart. *Source:* Adapted from *E-Train* (a publication of the National Environmental Training Center for Small Communities), 4(3), Fall 1995.

and the metal purified and reformed into a new product-processes requiring energy, other resources, and labor. Reusing the component in its entirety reduces these inputs. One common consumer product in which many components are reused is a motor vehicle. When you need, for example, to replace an alternator in your car, your "new" alternator is probably one that has been *remanufactured* from one that was previously discarded. To remanufacture an old alternator, it is taken apart, cleaned, inspected, and then rebuilt to specifications at least as good as the original. This saves about one-half of the energy and labor needed to manufacture a new alternator. Between 70% and 90% of all auto parts (examples are clutches and brake calipers) are remanufactured. Other products often remanufactured are used telephones, toner cartridges, office furniture, and even subway cars.

Recycling. "I think; therefore I pollute" is a recently coined phrase. No matter how successfully we pursue P^2 and reuse we will produce some waste. Fortunately, many materials can be recycled. In 1995, nearly 7,000 curbside programs representing 84% of American cities collected paper, steel and aluminum cans, glass containers, and, sometimes, plastic containers or waste oil. In states that charge deposits on beverage and other containers, recycling rates are especially high. Prior to the mid-1990s, recycling programs, with the exception of those for high-value goods like aluminum, often drained community resources by costing more than did landfilling or combustion. Some communities had to pay to have newsprint hauled away. That situation has changed, and even newspapers are often a source of profit. Paper recycling reached 40% in 1994, and more than 80% of the recycled paper is postconsumer waste, not just mill or printing plant scrap. The problem is now finding enough scrap paper to meet the demands of recycling mills. Research is even being conducted to find means to use paper from fast food establishments, paper that is often contaminated with food. In this case composting the recovered paper, not reusing the fiber, is the goal.

Although recycling programs are still sometimes a cost to a community, recycling eliminates the cost of landfilling wastes and provides economic and job opportunities through recycling industries. It also prevents environmental costs because it uses less energy and conserves resources. Although recycling does produce pollution, emissions are usually lower than those produced by the use of virgin resources. Recycling rates in the United States increased from 10% in 1986 to 22% in 1993 (review Figure 2.2). Details about specific recyclable materials follow:

- Aluminum: About two-thirds of aluminum cans are now recycled. Cans were successfully recycled before the market for other recyclable materials improved because aluminum is a valuable metal and cans are relatively easy to collect. However, the recycling of aluminum foil, pie pans, frozen dinner trays, lawn furniture, and the aluminum used in siding, gutters, and window and door frames has been less successful. Recycling aluminum uses 95% less energy than required to mine it from bauxite and reduces air and water pollution about 95%. It is also less costly to recycle aluminum. Another benefit is that only about 1% of the bauxite ore from which aluminum is recovered is found in the United States Thus, consumer aluminum discards are, in effect, an aluminum mine.
- Steel cans: Recycling steel cans (sometimes called tin cans because of a tin lining) saves 60%-70% of the energy needed to produce them from ore. In 1994, 53% of these cans were recycled.
- Glass: Recycling glass saves energy, although not as much as for metals. About a third of glass containers are recycled.
- Paper: 40% of paper was recycled in 1994 and paper manufacturers are working toward 50%. Above 50%, it becomes progressively more difficult to recover good-quality paper for recycling, although some may be composted and low value but clean paper can provide fuel to MSW combustors. About half of the paper currently recycled is from old corrugated containers.
- Plastic: Plastic wastes generated in the United States have increased about 10% a year for the past 20 years – a striking increase. Not only containers, but products like automobiles are using much more plastic. As compared

to 40 pounds of plastic used in a 1960 automobile, the figure in 1992 was 300 pounds. The value of used plastics also has rapidly increased. However, only a small percentage is currently recycled because of the difficulty of collecting and separating the many types of plastics. Recycling the polyethylene terephthalate (PET) found in soft drink bottles saves about 47% of the energy needed to make these from raw materials.

- Used oil: Of 1,378 million gallons of used oil generated in the United States in 1991, only 4.1% was refined into new automotive oil, as compared to 13.4% that was illegally dumped. Most of the rest was burned as fuel.
- Yard waste: There is not a market for yard waste, grass, leaves and brush, but communities often collect and compost them and the compost can be sold to improve soil quality. Homeowners often compost their own yard wastes.

See Table 8.2 for examples of other ways that recycling is used. Recycling has had many successes, but many challenges remain. With the exception of aluminum cans, more discarded materials are landfilled or combusted than are recycled. Or consider that more than 32 million used appliances – air conditioners, dishwashers, dryers, washing machines, refrigerators, freezers, stoves, microwave ovens, and water heaters – were thrown out by Americans in 1988. A number of states have banned appliances from their landfills. Appliances usually contain recyclable metal, but it is contaminated with other materials. Nonetheless, in one instance, a Connecticut electric utility collected discarded freezers and refrigerators, dismantled them, and recycled their components. To recover and purify metals from appliances produces pollution and requires energy. It is preferable to recover whole component parts from appliances as Xerox does with its copiers. To effectively do this, the product needs to have been designed for disassembly. Not surprisingly, recycling has its own environmental problems. Consider paper. It contains not only paper fiber, but coating, fillers, and pigments. These must be removed during recycling but are not currently reusable and must be landfilled. Recycling also has associated health and safety problems, such as those entailed in composting, land spreading sludge, or recycling newspapers. Workers have been badly injured or killed in accidents involving recycling equipment, and EPA is currently sponsoring a study to evaluate potential health effects and hazards of MSW recycling. Another problem is that recycling facilities have begun to suffer the same resistance to siting that plagues landfills and incinerators, the not-in-my-backyard, or NIMBY, phenomenon.

Society can promote recycling in a number of ways. One is technology development. Plastics contain a number of different resins, which, if mixed together, are useful for manufacturing only a few products. As technology develops to sort plastics efficiently and economically, each can be recycled to specific higher-value uses, or new technologies may find more uses for mixed

TABLE 8.2 Recycling is not just about paper and cans

Organic wastes	Yard waste and kitchen food waste can be [] a community composting program. Sewa[] animal waste, after treatment, can be use[] land as a soil amendment.
Old clothing and textiles	In the nineteenth century, cotton rags were rou[] paper. Old woolens were reprocessed into new clothing. M[] today recycle their clothes by selling them at yard sales or donating them to charity. Old clothing is also exported to third world countries. Clothing that is not reusable may be reprocessed into rags or mixed into asphalt to make roof shingles.
Grease	American fast food restaurants discard 2.4 billion pounds per year of used cooking grease. It can be processed into a poultry and cattle feed ingredient or used as a lubricating oil.
TV sets and telephones	European manufacturers of TV sets have formed a consortium with glass manufacturers to recycle picture tubes from TV sets. Germany's largest telephone company has formed a consortium to recycle telephones.
Building materials	Wood and other materials from old buildings can be recovered and reused. Builders are also incorporating novel materials into new construction — discarded tires, plastic bottles, metal cans, concrete (from incinerator ash), and insulation (from old newspapers). It is sometimes difficult to obtain building permits for these structures.
Toilets	California has a program to crush discarded toilets into an aggregate that is used in concrete for road building projects.
Telephone poles	Telephone poles produced in earlier years contained creosote and pentachlorophenol. However, these preservatives are now banned. So recycling wood that had been treated with them was not possible until Louisiana State University developed a *bioremediation* process. Poles are chipped into small pieces and composted using microorganisms that can degrade creosote and pentachlorophenol. The cleaned wood fiber is then sold to paper mills.
Fluorescent lamps	EPA encourages the use of fluorescent lamps because they are more energy efficient than incandescent bulbs. About a *half-billion* fluorescent lamps are used yearly in the United States. They contain a glass tube, aluminum end caps, inert gases and phosphor powders and 20 to 50 mg of mercury (depending on bulb size). Because of the mercury, some states treat fluorescent lamps as hazardous waste and ban them from MSW landfills. Both European and American businesses have developed techniques to recover and recycle the mercury and aluminum as well as the other components.
Radioactive materials	Rather than discard radioactive used pumps — especially as a longterm storage site for nuclear waste is not yet available — the nuclear power industry is recycling metal from old pumps into new pumps for use within the industry.

. Economic incentives can promote recycling. Government can pro-
subsidies to businesses that use secondary (used) rather than virgin mate-
ls. Social and behavioral changes can improve recycling rates. State and local governments can support recycling and waste reduction education in
schools; sponsor television, radio, or newspaper advertisements that promote
recycling; emphasize that it does matter whether individuals sort their trash or
buy products with recycled content. Political or governmental tools can enhance recycling, as when a law or regulation requires specific actions, such as
requiring government offices to purchase materials containing recycled content or requiring citizens to pay deposits on tires, car batteries, furniture, or
appliances at the time of purchase. For example, Idaho charges a dollar tax
on each new tire sold and uses the revenue to subsidize recycling programs. At
the federal level, President Clinton issued an executive order requiring federal agencies to purchase printing and writing paper containing at least 20%
recycled material by the end of 1994, and 30% by 1999. Businesses may voluntarily decide to make changes. L. L. Bean of Maine has teams working to identify means to reduce waste. EPA has a voluntary program, WasteWi$e, that asks
businesses to set waste-reduction goals and to provide a schedule as to how
they plan to meet those goals. They are also asked to buy or manufacture recy-

---

BOX 8.3

The fate of junked cars. About 75%, by weight, of a discarded automobile is reused or recycled. Much of the iron is recycled. Lead-acid batteries, starters, tires, generators, and other parts are typically reused or
parts of them recycled. The remainder of the car referred to as fluff (the
plastics, liquids, and glass) is usually landfilled. Germany has mandated
that auto manufacturers take responsibility for the vehicles that they
manufacture at the end of their useful life. American manufacturers selling cars in Germany must also comply. Three European companies,
BMW of Germany, Fiat of Italy, and Renault of France, have developed
dismantling, shredding, and recycling facilities for their discarded cars.
All three companies have banded together to share information and
technology, and each accepts used vehicles from the others. Another European car manufacturer, Volvo of Sweden, is working on an LCA of a
Volvo to evaluate the environmental impact of its manufacture, use, and
disposal. Results will be used to design more environmentally sound vehicles, which will include means to disassemble them at the end of their
lives and to reuse or recycle their components. See Chapter 12 for steps
American car manufacturers are taking.

cled products. As the name WasteWi$e implies waste reduction can provide financial rewards.

Treating MSW. A major means of treating MSW is combustion. Because the word incinerator has a poor reputation, the U.S. EPA prefers the word *combustor* and that usage is followed here. In 1993, about 16% of American MSW was combusted. Proponents of combustion see it as an important component of MSW management that is especially significant in large-population centers, where landfill space is limited. Combustion ash takes up only about a third as much landfill space as the MSW itself. Some believe that combustion also reduces the toxicity of solid waste because ash leachate has a lower solids content than leachate from raw MSW. Ash also does not have the organic materials or microorganisms that make garbage objectionable. Ash, unlike garbage, contains nothing that can generate the sometimes explosive methane gas. MSW combustors are highly regulated, but pollution concerns remain. Wastes containing the metal mercury are a special problem because this volatile metal is difficult to recover completely from combustor stack gases. Other metal products burned in combustors – batteries, and electronic and other consumer products – also produce metallic and dioxin emissions. However, such emissions will be greatly reduced under stricter regulations that EPA is promulgating. Supporters of MSW combustion assert that under the new EPA standards, MSW combustors may burn as cleanly as natural gas.

Those opposed to combustion do not believe that it serves a useful purpose. They object to air pollution, especially that caused by heavy metals and dioxins. EPA's stricter regulations on these emissions may allay those concerns, but those objecting to combustion are also concerned about landfilling the ash produced. About 90% of combustor ash is landfilled. Removing organic materials by combustion leaves the metals in the ash more concentrated. Heavy metals sometimes leach from the ash. Leachate is water that has percolated through a landfill, dissolving contaminants as it goes. Although leachate is treated, there is still concern that it may contaminate drinking water. Lead, cadmium, and mercury are special concerns. Removing as much metal as possible from MSW could allay this concern about landfilling ash and, at the same time, reduce air emissions. Lead-acid batteries from motor vehicles are already prohibited from combustors in many states. It is harder to remove small button batteries, which have been a major source of mercury in MSW. Although increasing amounts of metals are removed from MSW before combustion, metals remain. However, there is currently research on technologies to stabilize metals within the ash. The ash can be vitrified (converted into a glasslike material), greatly reducing leaching.

Combustor ash is rich in salts. These are a concern for the same reason that piles of salt used to control road ice in winter are a concern. Salts dissolve in water and run off into surface water or percolate into groundwater. Thus, even ash free of metals must be landfilled in a way that prevents salts from leaching into the environment. There is increasing interest in using rather than landfilling ash. Less than 10% of American combustor ash is used, but several European countries use most of theirs. They first stabilize ash to prevent it from leaching metals and then incorporate it into road beds. Although the technique is not yet cost-effective, ash can also be incorporated into cement blocks. A number of other potential uses also exist. If traditional resources become more expensive in the future and better technologies develop, progressively more ash may be put to use rather than landfilled.

The amount of MSW combusted in the United States increased from 10% to 16% between 1986 and 1992 (see Figure 2.2). Combustion recovers some of the fuel value of plastics, paper, and other carbon-containing wastes by producing steam and electricity. To generate electric power by burning MSW effectively, a combustor needs a sufficient supply of fuel-rich materials such as paper and plastics. For this reason, some believe that combustion hinders recycling; however, if the value of recyclable materials continues to be strong, it is likely that increasing amounts of materials will be recycled. Only a small percentage of plastics is recycled. Plastics are made from petroleum, 90% of which is burned as fuel; only 4% is used to manufacture plastics. Some argue that, because we burn 90% anyway, why object to burning an additional 4%? Although combustion is the most common means of treating MSW, organic wastes can be handled by other methods. Grass, leaves, and other yard wastes can be composted, as can many food wastes. During composting, microorganisms degrade the organic materials to yield a soil-like product. Worms are also sometimes used to compost food wastes. These are forms of biotreatment. More broadly, biotreatment can be used to treat many wastes, hazardous and nonhazardous. See the description of bioremediation in the section, Hazardous Waste.

DISCUSSION QUESTIONS

The operator of an MSW combustor likes to show visitors an aquarium with fish living in untreated leachate from combustor ash. The leachate is salty, but the fish seem to do well. For comparison, he shows a container of noxious-looking and ill-smelling leachate from MSW that has been landfilled without combustion, a medium in which fish obviously could not live.

(a) What conclusion would you reach from this demonstration?
(b) Is it important to maintain the option of combusting MSW? Why?

Landfilling. Landfilling MSW without combustion also has problems. Landfills are engineered to minimize water infiltration, but they nonetheless produce leachate, which must be collected and treated. Large facilities, like New York City's Fresh Kills Landfill, produce enormous amounts of leachate. Especially in the years immediately after waste is landfilled, leachate contains foul-smelling organic compounds attractive to the growth of many microorganisms. This does not usually pose a problem because the leachate is collected and treated to meet EPA standards before it is discharged to a waterway. Landfills that contain carbon materials produce methane gas, which must be properly managed to prevent explosions. However, methane is the major component of natural gas and an increasing number of landfills collect this gas for use as fuel.

Historically, trash was disposed of in open dumps, often located close to rivers or over groundwater. In fact, in some parts of the United States, dumps were sometimes deliberately placed on river banks so spring floods would wash the trash away. Open burning of trash in dumps continued well into the 1970s. Starting in the 1980s, strict regulations as to where landfills could be sited and how they had to be built, maintained, and monitored came into force. When possible, landfills are sited over near-impermeable clay that provides natural containment for the waste. To inhibit leakage of the water percolating through them further, landfills have a synthetic liner and a collection system to recover and treat the leachate. Monitoring wells are built around the landfill and regularly sampled to detect leaks. Built according to strict regulations and carefully engineered standards, they are clearly safer than older landfills or dumps.

An ongoing concern about any landfill, even one lined underneath and on top with state-of-the-art liners, is that it may leak before the waste has biodegraded to the point where leakage is of much less concern. New landfills address this problem by using layers of clay and bentonite soils along with geomembranes, which are specially engineered synthetic membranes underneath a landfill. A similar layering of special soils and geomembranes is used to cap a full landfill. A facility carefully constructed to specifications may remain leak-free for many years, past the point where leakage is a major concern. However, sophisticated landfills are very expensive to build, and thus provide an additional waste-reduction incentive. The percentage of MSW that is landfilled is already decreasing: In 1985, about 83% of MSW was land-disposed; this figure was reduced to 62% in 1993 (see Figure 2.2).

About half of the 6,000 landfills in use in 1991 may be closed by 2000. The United States is a large country, which has space for more landfills. However, NIMBY is a potent force that often prevents the siting of these facilities. There are a number of reasons for the NIMBY phenomenon. For a modern landfill

BOX 8.4

Speeding up the rate of degradation in landfills. A major problem with a typical landfill is that, although organic materials within it can biodegrade, they do so extremely slowly. This is true because landfill microorganisms lack the water and nutrients that would speed the breakdown process. To address this problem, the use of landfills as bioreactors is being tested. Instead of treating and releasing collected landfill leachate to a waterway, operators circulate the leachate back into the landfill. Leachate is rich in organic chemicals, and, kept within the landfill, these provide an energy source to the microorganisms that degrade organic wastes. Recirculating leachate also greatly hastens the stabilization of landfill wastes. A landfill is considered stabilized after most of the organic material has biodegraded. The results are much lower amounts of organic contaminants in the leachate and much less generation of methane gas. In an ordinary landfill, reactions leading to stabilization can take many years; however, in a landfill that recirculates leachate, this can happen much more quickly. Furthermore, during the years that it is active, the bioreactor landfill produces much larger quantities of methane gas, which can be captured and used as fuel, than an ordinary landfill. Currently a landfill, once full, is closed by capping it with an impermeable clay layer and then monitoring its discharges for 30 years to warn of releases of potentially harmful materials. However, if the waste is stabilized by the process described here before closure, it may be possible to reduce monitoring to as little as 10 years. By 1994, eight United States landfills had been built to test the bioreactor concept, four more were being constructed, and another 4 were being planned.

to be cost-effective, it must be large. But a small community does not need a large landfill and cannot afford to build one. Instead, regional landfills serving many communities are built, with social costs. People have no sense of ownership in a landfill many miles distant. People in the community hosting the landfill feel resentful at taking others' trash. There are also noise and sometimes road degradation and pollution associated with heavy truck traffic, and property values are often adversely affected. Also, landfills and other waste facilities have been disproportionately built in poor and minority communities. As these communities have become more aware of the practice, they have increasingly begun to object to it. Another social objection heard regarding landfills is that we are leaving our mummified trash for future generations to manage.

We can reduce waste, but we cannot eliminate it. Materials can be reused or recycled, but except for metals, which can be recycled indefinitely, recycled materials also become trash and will need to be combusted or land-disposed.

Some materials cannot be recycled at all, and, although combustion lowers waste volume, there is still ash to be landfilled. Thus, waste disposal will remain with us into the foreseeable future. However, as landfills continue to evolve, they may become ongoing operations rather than just places to be filled, capped. and left. A stabilized landfill may be mined for the recyclable or reusable materials that it contains. Mining also provides space to landfill more material. Or, in some locations, closed landfills are developed as golf courses, parks, or additional space for wildlife.

Batteries and the waste management hierarchy. Batteries are a product that we consume in huge quantities. Each of the billions of disposable batteries that Americans throw away each year takes about 50 times more energy to produce than it provides when used. As throw-away items, batteries use resources inefficiently and contribute metals, including heavy metals, to MSW, complicating both combustion and landfilling. Can batteries be better managed?

- P^2 means avoiding batteries when possible. Buy toys and appliances that don't need them. When purchasing batteries, buy rechargeable ones. These take energy to recharge, but less than required to manufacture them from virgin materials. In regions with plentiful sunlight, rechargeable solar batteries can be used. Button batteries, used in watches, calculators, hearing aids, and hospital and military equipment, sometimes contain up to 40% by weight of mercury or, sometimes, silver. These have been the major source of mercury in MSW combustor emissions. A mercury-free button zinc air battery is now available.

BOX 8.5

Landfill mining. In landfill mining, materials are removed either because they have value or because space is required for more waste. Thousands of American landfills will be full within the next few years, and these must be closed if a way is not found to extend their lives. Many older landfills contain up to 70% dirt and only 30% trash. In Massachusetts, the Read Company developed equipment to excavate trash and to feed it onto a machine, which separates bulky items from dirt and smaller items. This company anticipates that, even if recovered materials have little value, just removing the dirt creates valuable space to landfill more trash. The dirt removed can be used as a dump cover. Germany has also begun landfill mining. Recovered materials are taken to a sorting plant, where wood, rubble, and metals are removed. Materials not worth recovering are incinerated or returned to the landfill.

- Button batteries are valuable and can be recycled. Minnesota has started a button battery collection program, as have some European countries. Some jewelry stores also encourage their return. The rechargeable nickel-cadmium (Nicad) batteries, used in appliances, are also recyclable. Eventually they are discarded, to become the major source of nickel and cadmium in MSW. However, Nicad batteries can be easily removed from the newer appliances on the market, allowing them to be handled separately. Some states propose recovering them from the waste stream. American manufacturers do not consider household batteries (A,C, D, and 9V) worth recycling, although the Swiss voluntarily collect and recycle about 60% of them. Some American communities collect them to divert the metals away from MSW combustors and landfills. A consortium of American, European, and Japanese battery manufacturers is now examining means to recycle A, B, D, and 9V batteries economically.

DISCUSSION QUESTIONS

Consider how you use household batteries, especially throw-away ones.

(a) Could you use fewer of these without inconveniencing yourself?

(b) How could you reduce the environmental impact of the batteries that you do use?

(c) Repeat this exercise with another product of your choice.

HAZARDOUS WASTE

In earlier years, hazardous waste was disposed of in lagoons or trenches or injected into deep underground wells. Some was released into waterways with little or no treatment or was disposed of along with MSW. At disposal sites, corrosive wastes sometimes ate through their metal containers and escaped into the environment; volatile wastes escaped their containers and evaporated into the air; liquid wastes percolated through the soil into groundwater or ran off into surface water. When the companies that generated or disposed of hazardous wastes changed hands or went out of business, the location or even the existence of some sites was forgotten. In earlier periods, starting in the nineteenth century, many municipalities produced town gas from coal or oil for use in lighting. Many of these are modern hazardous waste sites. Other old sites include nineteenth- and twentieth-century mines, highly contaminated with heavy metals. Casual disposal of hazardous waste was not ordinarily malicious. All kinds of waste had always been disposed of casually. So long as the population was small and the industrial and military operations that produced waste were small, waste sites were largely disregarded. But industrial and military operations grew in number and size. At the same time, population grew and cities spread out over the countryside. It was inevitable that people and hazardous waste sites would make contact.

What are the wastes of concern, and why are they of concern?

Hazardous waste is not a specific chemical. It is a legal term given to waste that has one or more of the following characteristics – it is ignitable, corrosive, reactive, or toxic. A *toxic* substance is one that can cause adverse effects to living organisms exposed to it; examples are arsenic and cyanide, pesticides and heavy metals. A *corrosive* substance causes injury at the point of contact: the skin, eyes, lungs, or mouth. Strong acids and alkalis are corrosive, as are chlorine and hydrogen peroxide. The concentration of hydrogen peroxide found in household products is not corrosive, but concentrations used in industrial and laboratory settings can be much higher. An *ignitable* or *flammable* substance is one that can catch on fire; it is a fire hazard. Petroleum distillates or many organic solvents are ignitable. A *reactive* substance reacts violently with water, air, or other substances. An example is old containers of ethers, used in industry and laboratories, that form explosive peroxides. Familiar reactive substances are dynamite, gun ammunition, and firecrackers. Reactive substances in the household include chlorine bleach and ammonia; mixed together, they are dangerously reactive.

In addition to regulating *characteristic wastes*, which have one or more of the properties described, EPA regulates *listed wastes,* such as the sludge produced by a metal-plating operation. Although hazardous wastes are referred to as solid wastes under RCRA, solid wastes are sometimes really liquids or containerized gases. In 1987, the United States generated about 274 million metric tons of hazardous wastes, greater than half of the total generated worldwide. Unless hazardous wastes have escaped into the environment, they are of most concern to the workers who handle them. Ignitable or reactive wastes can threaten materials as well as people, as when a corrosive reacts with metal or other material surfaces, causing them to wear away.

What are the sources of hazardous waste?

Most hazardous waste is generated by large facilities that manufacture chemical, petroleum, metal, plastic, and textile products. Military operations also generate large quantities of hazardous waste, and, to varying extents, so do other government agencies. Hospitals, many universities, and commercial laboratories generate small amounts of hazardous wastes as do small businesses, such as gas stations, photographic developers, beauty parlors, and dry cleaners. Some wastes are exempted by law, including certain mining wastes and coal combustion ash. Household hazardous wastes, which constitutes about 1% of total hazardous wastes, are exempt from the law even when they contain exactly the same chemicals that are regulated when produced by industry. Waste paint is the household hazardous waste produced in the largest amount; others are paint thinners and strippers, pesticides, cleaners and polishes that contain petroleum distillates, and alkaline drain and oven cleaners. Even some discarded cosmetics, such as acetone nail-polish remover are hazardous waste. Another large-volume hazardous waste familiar to those who change their own car oil is used oil. Many do-it-yourselfers dispose of waste oil, and EPA estimates that about 150 million gallons a year ends up in landfills or sewer systems. Additional quantities are poured on the ground. Used oil is both ignitable and toxic. EPA could regulate it as hazardous waste, but with millions of do-it-yourselfers, this is not practical. Instead, EPA developed standards for commercial handlers of used oil: If the oil is destined for disposal, it may have to follow hazardous waste regulations, but if it is to be recycled, it is treated less strictly. MSW combustor ash, because it was produced by household trash, was previously exempted from treatment as a hazardous waste. However, under a 1994 court ruling, the ash must be tested to assure that it passes an EPA-specified leaching test. If it does not pass this test, it must be handled as a hazardous waste.

How are hazardous wastes controlled?

Since RCRA was passed in 1976, hazardous waste has been regulated cradle to grave:

- If a hazardous waste leaves the facility generating it, an identifying document called a manifest must go with it.
- If the waste goes to a storage facility, the storage facility receives a copy of the manifest.
- When the waste goes to a treatment facility, that facility also receives a copy.
- Finally, the disposal facility receives a copy and sends a copy back to the generator. The generator then sends a copy to EPA or a state agency. This procedure assures that a paper trail documents the waste every step of the way, cradle to grave.

Hazardous wastes and the waste management hierarchy

- P^2: The manifest system is a control, not P^2. However, the system stimulates P^2 because companies, seeking to avoid the costly regulations, treatments, and legal liabilities associated with hazardous waste, work to reduce the amount generated; review Tables 2.1 and 2.2 for examples of P² in industrial settings.

- Recycling and reuse: Many industrial hazardous wastes can be recycled. For example, hazardous organic solvents are commonly used in industry; after use, solvents can be purified and reused, a process that can be repeated many times. Indeed, many consider this P² rather than recycling. However, some waste is necessarily produced because the contaminants removed during solvent purification are a waste. Sometimes a company produces a waste – a by-product – that may be useful to another company. These by-products are sometimes advertised in periodicals with the intent of finding buyers for them.

- Treatment: The type of treatment used to reduce waste volume or toxicity depends upon the nature of the waste. Treatments may be physical, chemical, thermal or biological (see Table 8.3). A biological method, bioremediation, uses microorganisms to treat waste. Several ways that microorganisms can be usefully applied have already been described, including degrading the organic matter in wastewater or in landfills. The use of microorganisms

TABLE 8.3 Hazardous waste treatments

TREATMENT	EXAMPLES
Thermal	Combustion is a thermal treatment to destroy organic substances. Metals and salts remain in the ash, which must be disposed. In stabilization, a metal-containing waste is treated with heat to solidify it into a hard mass. Metals are difficult to leach from this solidified mass.
Chemical	Neutralization converts an alkaline or acidic substance to a neutral form. An acid is neutralized with an alkali, or an alkali with an acid. Precipitation is a chemical treatment used for a solution of metals. A substance is added that causes the metals to become insoluble and to precipitate from solution. The precipitate is collected for further treatment.
Physical	In filtration, a liquid is separated from a solid by a membrane. The liquid passes through membrane pores while solid particles remain on its surface. Distillation is sometimes used to separate a mixture of liquids. The mixture is heated to drive off the more volatile substance, leaving the chemicals with higher boiling points behind.
Biological	Microorganisms can degrade certain hazardous wastes. Typically, they use the wastes as nutrients degrading them at the same time. Certain plants are able to concentrate heavy metals from soil or water without harm to themselves. These are being explored as a means to remediate contaminated sites.

to treat hazardous wastes is being intensively studied, and more applications regularly developed. Microbes can, for example, be grown on filters and can degrade 95%-99% of certain waste organic gases that are passed over them. A beneficial characteristic of bioremediation is that it works with natural systems. Hazardous waste treatments are discussed further later in relation to hazardous waste site cleanup.

• Disposal: At the bottom of the waste-management hierarchy is disposal. Unlike MSW, a hazardous waste cannot legally be landfilled until it has been treated to destroy its toxicity or else stabilized to prevent its hazardous components from leaching. More than one treatment may be necessary to accomplish this. Even then, some hazardous wastes may not pass an EPA leaching test; that is, they still contain hazardous substances that potentially could leach from the waste. Such wastes must be disposed of in a hazardous waste landfill and must be in the form of solids, not liquids. Some waste is still disposed of by injecting it into deep underground wells, but complex and costly regulations and potential liability greatly limit such disposal.

DISCUSSION QUESTIONS

1. Hazardous wastes have sometimes been treated by spreading them on land set aside for this purpose and allowing natural processes to degrade them.
   (a) What natural factors contribute to degrading organic waste?
   (b) What are potential environmental health and safety shortcomings of such land spreading?
   (c) What wastes cannot or should not be treated in this manner?
   (d) Consider hazardous chemicals that are buried in the sediments of a water body. What natural factors, available above ground, are no longer available to assist in chemical degradation?

2. Bioremediation has great potential.
   (a) Do you envision potential risks of this method?
   (b) If bioengineered organisms are used, could they increase the risk?

HAZARDOUS WASTE SITES

A hazardous waste site is one where uncontrolled release of hazardous chemicals either has occurred or is likely to occur. In the late 1970s, Love Canal became a notorious example of people and hazardous waste coming into contact (see Box 8.6). More than 35,000 hazardous waste sites have been identified around the United States. EPA decided that 22,000 of these did not constitute enough of a threat to require federal action, although many will be cleaned up by states or site owners. Sites that EPA identifies as posing a high risk are placed on the National Priority List (NPL); these are referred to as Superfund sites. Love Canal is one of about 1,300 Superfund sites and more continue to be added. Of the NPL sites, 2% are considered urgent health hazards and another 35% health hazards. Other than urgent hazards, which are handled very quickly, Superfund cleanups have been very slow. Cleanup is funded in two ways. Some money comes from a tax paid by oil, chemical, and manufacturing industries, but this money is used only when the parties responsible for contaminating the site cannot be found or are bankrupt; sites for which no responsible parties are available include some dating from the nineteenth century. Otherwise, those who created the site are responsible for its cleanup. If that company has changed hands, its buyer becomes responsible for the cleanup. For example, Hooker Chemical disposed of waste in Love Canal. However, it was Occidental Chemical, the company that bought Hooker Chemical, that had to assume a significant portion of the responsibility for the remediation of Love Canal. Another well-known hazardous waste site, Valley of the Drums, is shown in Figure 8.4.

In addition to the 35,000 sites identified by EPA, there are many sites owned by the federal government; the DOE alone is responsible for about 20,000. Many of these contain radioactive wastes from former defense operations, often mixed with other hazardous wastes. Cleanup estimates for DOE sites are staggering, ranging between $500 billion and $1 trillion over a period of 30 years or more. Since it is unlikely that all can be cleaned up, priorities will have to be set. The highest-priority sites are those presenting risks to people living near them.

BOX 8.6

Love Canal. The canal was built in the early twentieth century but never used. Later, from 1942 to 1953, Hooker Chemical Company disposed of 22,000 tons of hazardous waste in the canal, mostly its own waste, although about 25% reportedly came from the United States Army. The company used methods considered standard practice at the time. In 1953, Hooker filled in the site and capped it with a layer of protective clay. At the request of the city of Niagara Falls, it sold the property to the Board of Education for one dollar. A stipulation in the deed warned that hazardous chemicals were buried at the site, but the city later blamed Hooker for not informing it of the nature of the wastes and their health hazards. Occidental Chemical purchased Hooker in 1968 before the problems of Love Canal became known, and it is Occidental that has been held responsible for paying a significant portion of the cleanup costs.

BOX 8.6 *(continued)*

In 1954, shortly after the School Board bought the property, two schools were built, one next to the landfill and one directly over it. The board sold another portion of the site to a developer, who put up hundreds of houses, some directly around the landfill. In the 1960s, an expressway bordering the development was built. During this activity, roads and sewers were constructed directly over and through the site and the clay cap protecting the site was partially excavated, allowing rainwater to seep into the waste. The new expressway created the greatest problem: It blocked the normal path of groundwater migration through the area; the topographical characteristics of the site did not allow the groundwater to find an alternate pathway in which to flow. As a result, the waste and an increasing amount of groundwater and rainwater were trapped in a clay "bathtub" (the canal), which finally overflowed – along with its contaminants – into the basements and back yards of local homeowners. People began to report nauseating smells and some became ill. In the late 1970s, the federal government declared an emergency, bought several hundred homes near the site, and evacuated their occupants. The homes closest to the canal were destroyed and the rest sealed off. As of 1993, after more than 10 years of effort, about $325 million dollars had been spent just to contain, not remediate the waste. Above ground, the 40-acre site is surrounded by an 8-foot high fence. Below ground it is surrounded by a barrier drain system that directs leachate to a treatment system. The top is sealed with a thick layer of clay plus a high-density plastic membrane. Its bottom rests on low-permeability clay, greatly retarding downward movement of leachate. These measures are not a permanent solution, and the site will be maintained on an ongoing basis for many years. Occidental Chemical has assumed responsibility for maintaining the landfill's leachate treatment system.

New York state and the U.S. EPA decided in 1988 to redevelop the area and to sell, at reduced prices, the homes that had not been destroyed. Although this effort has been controversial, many houses have been sold. The present and former inhabitants are of two minds: Some, especially those who refused to leave their homes in the first place, believe the risks were greatly overstated and blame the city and state for disrupting the canal. Others remain angry and believe the area still poses unacceptable risks. There are also many court cases pending from people who say the contamination made them ill. There is no way to measure the psychological effects on the people involved. According to a resident in 1995, "Niagara Falls is no longer the honeymoon capital but the toxic waste capital."

FIGURE 8.4. Valley of the drums. *Source: The Courier-Journal*, Louisville, Kentucky. Reproduced with permission.

Some hazardous waste sites were created, not by the deliberate disposal of hazardous materials, but by leaking underground storage tanks. Because above-ground tanks were fire hazards, underground tanks became a popular means to store petroleum and other flammable chemicals. But over years of burial, the tanks corroded and leaked, contaminating surrounding soil and, sometimes, water. It is estimated that there are more than 2 million underground tanks in the United States located beneath gas stations and industrial operations. EPA is working with states to remove leaking tanks and to clean up the contamination they caused. To prevent future leaks, EPA now requires that underground tanks meet specific design, construction, and installation requirements and have leak detection and leak prevention systems.

DISCUSSION QUESTIONS

Think about Love Canal.

(a) In your opinion, did the city of Niagara Falls bear any responsibility for what happened?
(b) Should an industry that did not place the waste at a hazardous waste site be held liable for the cleanup because it now owns the business that was responsible?
(c) If not, who should be held responsible?

What are the hazardous wastes of concern, and why are they of concern?

Among the 10 most commonly found chemicals at hazardous waste sites are metals (lead, arsenic, mercury, and cadmium) and VOCs (benzene, vinyl chloride, trichloroethylene, and chloroform). The remaining 2 chemicals among the top 10 are PCBs and benzo[a]pyrene, which is a PAH. The health and environmental risks associated with these 10 chemicals are covered elsewhere in this book. The Agency of Toxic Substances and Disease Registry (ATSDR) has identified more than 2,000 chemicals at NPL sites. (The purpose of ATSDR, which was created by the Superfund law, is to deal with the risks posed by hazardous waste sites.) Each of the 2,000 chemicals has at least one hazardous waste characteristic; that is, it is toxic, corrosive, ignitable, or reactive.

Human exposure may occur through air, drinking water, food, and soil. The most likely exposure route for those living near sites is the water that may be used for drinking. ATSDR estimates that 20 to 40 million Americans live within 4 miles of the 35,000 hazardous waste sites identified by EPA and about 4 million live within a mile. Epidemiologic studies of these people indicated they may have a small to moderate increased risk of cancer and birth defects. Some studies showed no adverse health effects, but an ATSDR official noted that, taken together, the results were troubling. However, all these studies were based on proximity to hazardous waste sites, not on known exposure to chemicals. Current efforts are directed to finding individuals living near Superfund sites who were actually exposed. This can be determined by monitoring urine, blood, and tissue samples for the presence of chemicals found at those sites.

How is the risk of hazardous waste sites being reduced?

A potential Superfund site is first analyzed to identify the chemicals present, the type of soil, the way water drains from it, and possible water contamination. Other questions that need to be answered include, How close is the site to human dwellings? Are there routes by which people could be exposed to the waste (air, water, or food)? What are the best cleanup alternatives? If the site is determined to be among the 2% posing an urgent (imminent) hazard, this is dealt with quickly to protect workers and any nearby communities; an example of an imminent hazard is the presence of explosives. EPA analyzes the initial findings and, in consultation with others, selects a cleanup remedy from among the alternatives identified. The remedy must protect human health and the environment and also comply with federal, state, and local laws. An engineering cleanup design is then prepared. Next, the construction necessary to accomplish the cleanup is done; finally, the selected remedy is carried out. After cleanup, monitoring must be carried out for a number of years. If groundwater is being treated, cleanup could take many more years. Not including followup monitoring, the process may require 10 years or more. EPA is responsible for Superfund sites, although cleanups may be performed by others. At all stages, the public has a right to comment, although many believe their input has not always been seriously heeded in the past.

Examples of cleanup methods. There is an overlap between hazardous waste treatments (see Table 8.3) and techniques used to clean up hazardous waste sites. Site cleanup can, however, be much more complex than treatment of well-understood wastes within the confines of an industrial facility. Although 10 chemicals are most commonly found at Superfund sites, there are altogether about 2,000 hazardous chemicals; there are thus many potential cleanup remedies. Two examples of remedies, one primarily used for metals and the other primarily for organic chemicals, are the following:

- Stabilizing metals on site: Soil contaminated with lead or other heavy metals is often found at hazardous waste sites, especially old mining sites. The soil can be excavated and buried elsewhere in a landfill dedicated to hazardous waste. However, this process is expensive and the hazardous waste, although now at a safer site, remains hazardous. A second alternative is to treat the contaminated soil on site by stabilizing the metals. To do this, heat can be used to solidify the soil into a hard mass, or cement can be added. Stabilized metals are much less likely to leach.

In a new method, the TerraMet process, soil is cleansed of lead by the following procedure:

1. The soil is separated into its components (gravel, sand, silt, and clay).
2. The gravel is washed until most of the lead has been removed and then returned to the site.
3. The sand is shaken in water. Because lead is denser than sand, it settles out of the sand into the water and can be removed. The sand itself is then treated with a proprietary chemical solution that dissolves remaining lead. The clean sand is then returned to the site. The proprietary solution containing the lead is treated to precipitate the lead, which can be recovered.
4. Silt and clay are treated by a method similar to that used for sand and are also returned to the site.

The TerraMet process cleanses the soil and, at the same time, recovers the lead in a form allowing it to be recycled into lead products.

- Bioremediation: Living organisms can be used to remediate (bioremediate) chemicals at a hazardous waste site, at an oil spill, or in contaminated water or sediments. The waste site is first investigated in detail to identify the pollutants it contains and the characteristics of the water and soil and to identify what microorganisms are already naturally present. Most often, humans assist these naturally occurring microbes, which have already slowly begun to degrade the waste. To speed up the action of the microbes, nitrogen or other nutrients may be added to the soil. Aeration may be used if the microbes need oxygen. Perhaps the best known use of bioremediation followed the *Exxon Valdez* Alaskan oil spill. A physical cleanup method used was high-pressure water to wash oil from rocks. This was not effective and may have been harmful. But bioremediation helped: Naturally present microbes had already begun to degrade hydrocarbons in the oil along the shore and tidal flux provided needed oxygen. A nitrogen fertilizer was added along the coast to speed up the microbial action. Microorganisms that received fertilizer degraded both surface and subsurface oil three to five times faster than did microbes in unfertilized areas. Oil that could have taken 10 years or longer to naturally degrade was broken down in 2 to 3 years. Microbes were unable to degrade hydrocarbons that had already become immobilized in an insoluble asphaltlike material. However, this oil was not expected to affect biological systems adversely.

In the United States, bioremediation is primarily being researched and used as a cleanup tool for hazardous waste sites, oil-contaminated soil, or soil contaminated with munitions. Bioremediation can even be useful in metal-contaminated sites. Metals cannot be destroyed, but microbes can sometimes convert them to forms that are more easily dealt with or can carry out reactions that indirectly aid metal stabilization. For example, at one site, microbes converted sulfate to hydrogen sulfide. The sulfide reacted with heavy metal contaminants to produce insoluble metal sulfides, which were immobilized in place. This immobilization is analogous to thermal stabilization of metals in soil but is cheaper and more environmentally benign. Most often, natural microbes are used in remediation projects, but bioengineered organisms that can degrade specific chemicals are being developed. A *bioengineered organism* is one into which one or more genes have been introduced to make it capable of carrying out a task, which it previously was unable to accomplish. Recently, bacteria have been developed that can decompose dioxins and PCBs. Even in spots protected from the destructive power of sunlight, natural microorganisms do slowly degrade PCBs and dioxins, but the challenge is to destroy them rapidly with the bioengineered organisms. To date, the bioengineered microbes have only been tested in the laboratory. Their effectiveness must next be shown at actual waste sites, and it will also be necessary to show that they themselves present no risk.

Plants can be used in bioremediation (phytoremediation) and show particular promise of being able to remediate sites contaminated with heavy metals. Like animals, most plants can be poisoned by heavy metals. However, certain plants can accumulate as much as 40% of their weight as heavy metals without harm to themselves. These plants can then be harvested and burned to recover the metals. Plants may also prove effective in treating other chemicals. One aquatic plant took water contaminated with the explosive TNT from a concentration of 128 parts per million down to 10 ppm. A fascinating recent study demonstrated that sunflowers whose roots were submerged in water concentrated radioactive wastes from the water. The uranium concentration in the water, initially 100–400 ppb, was reduced enough in 24 hours to meet EPA's groundwater standard. Because the radioactive chemicals accumulate only in the roots, only they require special disposal. Although there have been field tests of this method, it has not yet been tested in a real world situation. The development of bioremediation, whether with microbes or plants, is in its infancy, but its potential is tremendous. It is less costly than technologies such as excavating and incinerating contaminated soil and, although assisted by humans, works with natural systems. Bioremediation does raise a familiar problem – it does not destroy the last small percentage of a contaminant. Microorganisms most often degrade a chemical because they are using it as an energy source to promote their own growth; as the concentration of the chemical reaches low levels, it can no longer support microbial growth, and when growth stops or slows, so does accelerated degradation of the chemical.

BOX 8.7

International use of bioremediation. Bioremediation has been primarily studied in the United States as a means to to clean up hazardous waste sites. However, its potential goes far beyond waste sites. European countries are beginning to use microorganisms to treat both air and water industrial waste effluents. In Austria, one large scale-project is using microbes to degrade the organic material found in MSW with the intent of eliminating the need to landfill the waste. Japan is exploring very ambitious uses of bioremediation such as using microorganisms to manufacture advanced biodegradable polymers or to produce clean-burning fuels like hydrogen. It is also studying a material produced by a bacterium that can absorb 1,000-fold its own weight in water. The hope is to use this absorbent material to remediate desert formation in the many world locales where that is a problem. Japan is also looking for ways that living organisms can be used to remove the greenhouse gas carbon dioxide from the atmosphere or to recover it from power plant stacks and convert it into substances that will not enter the atmosphere. Considering their visionary goals, it is not surprising that Japan's proposed uses for bioremediation are far from fruition.

DISCUSSION QUESTIONS

Consider the methods used to remediate soil contaminated with lead described in this chapter: soil excavation and removal, thermal stabilization, and the TerraMet process.

(a) What are some of the environmental impacts of each of these?
(b) Overall, which do you see as preferable? Explain.

Problems with Superfund. In the early stages of analyzing a hazardous waste site, attempts are made to identify parties responsible for contaminating the site, and these potentially responsible parties (PRPs) are then notified. Superfund operates on the basis of joint and several liability. This means that even if a company contributed only a small amount of contamination to the site, it could be required to pay all cleanup costs. In practice, a company will sue other responsible parties to obtain a fair portion of those costs. A PRP may not have been the company that actually contaminated the site but may have bought it at a later date; nonetheless, it is liable (see the description of Love Canal earlier in this chapter). This process leads to many, often lengthy and expensive, lawsuits. Superfund was passed in 1980; as of 1994, $13.5 billion had been spent, of which about 25% went to lawyers, not to cleanup. At the same time, only about 200 of the 1,300 identified Superfund sites were cleaned up.

Another source of controversy is that the law requires that a site be cleaned to its preindustrial condition. Attaining pristine cleanup standards at all Superfund sites could cost $300 billion or more and require several decades of effort. Many believe that cleanups should be taken to a 90–95% rather than 100% level because it is the removal of the last small percentage of contaminant that causes the major problem. They argue that this standard would allow cleanups to be completed much more rapidly and economically. All agree that hazardous waste sites located in neighborhood areas must be cleaned to meet strict standards. However, if the location is to continue as an industrial site or be paved over to provide a parking lot, the argument is made that a return to pristine conditions is not necessary. As A. J. Hoffman, author of the article discussed earlier, stated, "More Superfund effort should be spent on understanding the sociological reality of having trace contaminants in our presence rather than on the technological fantasy of being able to eliminate them." Some are unconvinced by these arguments and believe that cleanup standards should be more, not less, protective of health and the environment. Congress will be modifying the Superfund law in the near future. When it does, EPA wants to do more to protect the environment as well as human

---

BOX 8.8

A different cleanup standard. Among other reasons that businesses do not purchase land within a city to locate their facilities is concern that they must clean up the land to pristine conditions. So companies prefer to purchase open land, greenfields. Thus, not only is abandoned city land not redeveloped, but greenfield development leads to progressively less undisturbed land. For some years, cities wanting to redevelop abandoned urban sites have been pointing out that strict cleanup standards for industrial sites that are intended to continue as industrial sites serve no good purpose and inhibit land reuse. Inner-city Detroit has been particularly devastated by large abandoned tracts of land, some of which are said to look like a war zone. EPA has responded by funding pilot redevelopment projects at brownfield locations (abandoned industrial sites). When the Superfund law is reauthorized, it is expected that site-specific standards will allow a site to be cleaned up to a standard that depends upon its future use. It is also proposed that vegetation, especially native flora, be reintroduced to some brownfield sites to give urban areas some badly needed greenery and parks.

health. For example, if a metal is more toxic to animals than to humans, then standards should be set to protect them. This position is in keeping with the recommendation of EPA's SAB, which urged EPA to take ecological risks as seriously as human health risks.

DISCUSSION QUESTION

The strict liability provisions of the Superfund law have led to more responsible waste disposal, and also to P^2. In what ways might Superfund have stimulated P^2?

FURTHER READING

Solid waste

American Chemical Society. 1993. *Recycling, An Information Pamphlet.* Washington, D. C.: ACS.

1993. Life-Cycle Assessment: Tracking Impacts from Cradle-to-Grave. *Pollution Prevention News*, EPA 742-N-93–002, July-Aug., p. 9.

Shapiro, E. 1993. Pollution Prevention, Three Case Studies. *EPA Journal*, 19(3), 14–19, July-Sept.

Young, J. E. 1995. The Sudden New Strength of Recycling. *World Watch*, 8(4), 20–25, July/Aug.

Hazardous waste

1993. Health Effects of Hazardous Waste. *Health & Environment Digest*, 7(5), 1–8, Aug./Sept.

Graham, J. D., and Walker, K. D. 1994. How Clean Is Clean? *Health & Environment Digest*, 8(3), 17–19, Jun.

Graham, J. D., and Sadowitz, M. (1994). Superfund Reform: Reducing Risk through Community Choice. *Issues in Science and Technology*, X(4), 35–40, Summer.

Hoffman, A. J. 1995. An uneasy rebirth at Love Canal. *Environment*, 37(2), 4–9, 25–31, Mar.

Lewis, J. 1991. Superfund, RCRA, and UST: The Clean-Up Threesome. *EPA Journal*, 17(3), 7–14, July/Aug.

METALS

Chapters 5 through 8 of this text emphasized pollution from the perspective of the environmental medium affected, air, water or land. This approach can underemphasize important pollutants and partially mask the fact that many pollutants are multimedia, contaminating all media. So pollution will be approached from other perspectives in the remainder of this book. Chapters 9 through 11 emphasize specific pollutant categories: metals, pesticides, and the controversial, but important environmental estrogens. Another shortcoming of emphasizing one environmental medium at a time is the possibility of underemphasizing the fact that certain processes generate a great deal of pollution. Agriculture is one of these, as will be seen in Chapter 10. One highly polluting process, energy production and use, will be discussed in Chapter 12. Finally, Chapter 13 emphasizes pollution in a "medium" that intimately affects each of us: the home.

METAL POLLUTANTS

Metals are found naturally in soil and water, and enter food and drinking water even from uncontaminated soils. Natural levels of arsenic in water, for example, can pose health risks. But human activities have dramatically increased background levels. About 90% of all the cadmium, copper, lead, nickel, and zinc ever mined has been mined since the beginning of the twentieth century. A number of metals – iron, chromium, copper, cobalt, manganese, molybdenum, nickel, selenium, vanadium, and zinc – are essential to life; that is, they are nutrients. However, only small amounts are needed for good nutrition, and higher doses can have adverse effects (review Figure 3.2). Consider iron. Most people know iron is a nutrient; they may not know that a number of small children are fatally poisoned in the United States each year after eating their parents' iron supplement capsules. This chapter, after initially discussing metals generally, will move on to concerns specific to lead, cadmium, and mercury.

What are the pollutants of concern, and why are they of concern?

Three metals that pose special concerns are the heavy metals, lead, cadmium, and mercury. Anyone who has lifted a lead brick can appreciate the designation heavy metal, as compared to a light metal like aluminum. Other metals sometimes found at levels of concern in the environment include antimony, arsenic, cobalt, copper, manganese, molybdenum, nickel, selenium, vanadium, and zinc. Some environmentally persistent organic chemicals degrade very slowly in the environment, but they can degrade. However, metals are elemental: They do not break down. A metal's chemical aspects may change, but the element itself remains. Over years, metals in a given location may wash away, or the soil itself, carrying the metals, will erode and wash away.

Some concerns associated with metal pollution follow:

- Pollution in fresh waters is a concern because natural metal levels in these water bodies are often low and many lakes have small volumes or few outlets to displace polluting metals. Metal levels in marine waters are naturally higher than in fresh water.
- Pollution of coastal waters is a major concern. With 50% of our population living near a coast, these waters are increasingly contaminated by human activities.
- Older cities, which have traditionally had more metal-processing facilities than rural environments, are often highly polluted with metals.
- Metal contamination of agricultural soils presents serious problems. Many European soils, even some in Western Europe, are at or close to the top loading rate for some metals. Above the loading rate, the soil cannot safely be used to grow crops. Some agricultural villages in Eastern Europe have been deserted because of heavy metal contamination of soils. In Japan, nearly 10% of rice paddy land can no longer be used because of metal contamination.
- Some hazards are specific to areas affected by acid rain: Bacteria in acidified lakes can convert more elemental mercury to the highly toxic methyl mercury. Many metals – aluminum is a prominent example – are more soluble in acidified soils, and aluminum mobilized by acid can run off into nearby water bodies and adversely affect aquatic life.
- The heavy metals lead, cadmium, and mercury are 3 of the 10 most common pollutants found at hazardous waste sites. The metalloid arsenic is a fourth. Metal-contaminated soil can be remediated, but at heavy cost. Cleanup of large areas may be impossible.

Health effects. Metals have many potential adverse health effects, some of which will be discussed for lead, cadmium, and mercury. These metals are not nutrients, and they have a number of demonstrated adverse effects in humans, animals, and plants. One concern centering around these three metals is that living organisms concentrate them to levels much higher than found in the environment. Consider mercury, which algae very effectively recover and concentrate from water. The algae are eaten by small fish, which, in turn, are eaten by larger fish. At each step of this food chain, progressively higher amounts of methylmercury accumulate. Some game fish have methylmercury concentrations in their tissues 225,000 times those found in the water in which they live. Fish-eating birds and mammals can accumulate even higher levels. In an increasing number of locations, advisories have been issued to limit human consumption of fish because of high methylmercury levels.

People with poor nutrition or those eating large amounts of a contaminated food are most vulnerable to the adverse effects of metal exposure. Children, even those who are well-nourished, are more vulnerable than adults because they absorb more of the metals that they ingest across the intestine. Fortunately, some metals have opposing biological effects. Dietary calcium and iron inhibit lead absorption; selenium is protective against mercury, cadmium, and silver toxicity; zinc antagonizes some of cadmium's toxic effects. There are other examples as well, leading to the conclusion that a well-balanced diet with food from a variety of sources is an important safeguard against metal pollution.

BOX 9.1

Treating diseases with metals. The toxicity of a metal depends upon the dose to which a person or animal is exposed, the amount absorbed into the body, the person's nutritional and health status, and many other factors. A number of metals have been used beneficially to treat diseases. For many years before antibiotics became available, bismuth and arsenic were used to treat syphilis, although, not surprisingly, patients ingesting doses that were too high were poisoned. Metals still find medicinal uses. Bismuth, as in Pepto Bismol, is still used to treat some stomach problems. Gold finds use in the treatment of rheumatoid arthritis and lithium in treatment of manic depression. Zinc is used in some drugs. But metal remedies can be misused. So-called folk remedies containing lead, mercury, cadmium and arsenic are still employed in some places in the world. A number of poisoning cases have been reported among immigrants who have imported these "remedies" with them into the United States.

What are the sources of metal pollutants?

Smelting and refining of metal-containing ores or of metals recovered from the solid waste stream are major sources of metallic air emissions. Manufacturing of metal products is another. Air emissions may pose problems as air pollutants or after they settle onto land and water (see Figure 9.1). Mining operations that provide the ore to the metal processing operations generate large amounts of solid wastes. Rainwater or snow melt can leach metals from these wastes, which can run off to contaminate local water bodies.

In addition to the sources just noted, there are other pervasive and cumulatively important sources:

• Burning fossil fuels is one. Fossil fuels contain only trace levels of metals, but such huge amounts of coal and petroleum are burned that combustion is a significant source of metal emissions. Coal-burning plants also produce huge quantities of metal-containing ash, the disposal of which presents ongoing problems. Unless the ash is properly protected, the metals it contains can be leached out and run off to contaminate local water.

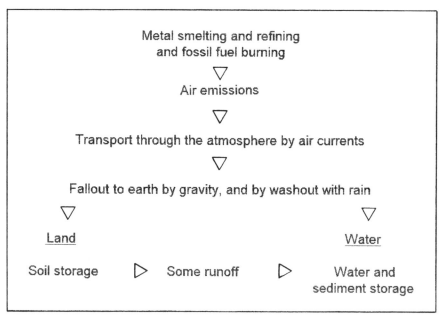

FIGURE 9.1. Transport of metals emitted to air. Metals are emitted as particulates, with the exception of mercury vapor. Metals bind to soil, especially if it is neutral or alkaline, although some slowly dissolve in water and run off into nearby water bodies. If the soil is acid, then a larger portion dissolves and runs off. When soil itself erodes and runs off into water, bound metals are also carried away. The amount of a specific metal stored in water, as compared to the amount in sediment, depends upon the chemical form of the metal.

- Another pervasive source of metal pollution is the metal-containing consumer products discarded at the end of their useful life.
- Industrial wastewater, sewage discharges, and urban runoff from streets and construction sites contain metals; even when levels are low, these sources can represent a continuing input of nondegradable pollutants.
- Some fertilizers and manure contain metals at levels above those found naturally in soils and contribute to the metal load of agricultural soils.
- Natural metal sources include volcanoes, forest fires, and sea salt sprays. Metals are also naturally present in the earth's crust. Although natural sources are important, it is human activities that are increasing the environmental load of metals.

The major route of human exposure to metals varies with the metal in question. With methylmercury, fish consumption is the major route of exposure; with cadmium, it can be shellfish for those who frequently eat them or it may be tobacco for smokers. For arsenic, food and drinking water may be significant routes of exposure. In the work place, air can be a significant source of metal exposure.

A summary of metal contamination of air, land, and water

- Air: With the exception of mercury, which is a liquid, metals are solids. Emitted to the air as particulate matter, they settle onto water and land. Relatively greater amounts settle out near emission sources, but metals can be carried long distances. Recall the example of industrial emissions from Eastern Europe and Asia carried to the Arctic, where they form a late-spring haze each year.

BOX 9.2

Ancient mining techniques. Although the majority of metal pollution has occurred in the twentieth century, humanity is fortunate that the ancient Greeks and Romans were unable to mine as much ore as their modern-day counterparts. Consider smelting, a process in which an ore containing a desired metal is heated and melted in order to separate out the metal. The ancient Romans smelted ores over open fires. The emissions were so noxious that they were noted in the writing of authors of the day; in fact, the pollution that resulted can still be detected around the world, recorded as high levels of lead and copper in ice, bogs, and aqueous sediments laid down 2,000 years ago. In the sixteenth century, large furnaces with tall stacks were developed. They polluted their surroundings badly, but not as much as the ancient method.

• Water: Point source discharges of metals include municipal and industrial wastewater treatment plants. Nonpoint sources include runoff from streets, construction sites, and metal-containing wastes disposed of on land. The atmosphere is a significant nonpoint source of metal pollution: At least two-thirds of the lead and mercury and more than 50% of other trace metals that enter the Great Lakes are from atmospheric deposition.

• Soil: Discarded mining residues containing high metal concentrations are significant local sources of soil contamination as are discarded metal products and combustor ash. Atmospheric fallout of airborne metal pollutants can be important, especially when sources are nearby. Some fertilizers, pesticides, and animal wastes applied to soil contain metals at levels above those naturally found. Metals originally discharged into water can also reach and contaminate soil; this can happen when sludge recovered from municipal wastewater is spread on land.

How are metal emissions being reduced?

The mandates of several laws control metal emissions. The discharge of many metals into water is controlled under the CWA. RCRA, which regulates solid wastes, places limits on how much metal is allowed to leach from solid wastes. Air emissions are controlled under the 1990 CAA amendments and will be more tightly controlled still when EPA's new regulations to reduce emissions of lead, cadmium, and mercury from MSW and medical waste combustors go into effect. As of 1996, there were no proposed restrictions on metal emissions from power plants, although some states do control them. If efforts to reduce VOC and criteria pollutant emissions from motor vehicles succeed, metal emissions will also be lowered because petroleum, like coal, contains trace metals. Another ongoing effort is a joint Canadian-American project directed to preventing further metal buildup in the Great Lakes. In a number of locations bordering coastal waters, efforts are under way to reduce metal discharges. Metal-processing operations are also being improved. A new smelter-refinery complex in Utah, using improved technology, is expected to be among the cleanest in the world and cost-effective as well. In addition to control steps, P^2 is being used, including replacing the heavy metal pigments used in the printing industry by vegetable-based ones and restricting permissible levels of heavy metals in both paper and plastic packaging.

DISCUSSION QUESTIONS

1. Are you concerned enough about the environmental impacts of metals to start removing metal-containing products from your

household waste before they are sent to a landfill or an MSW combustor? Explain.

2. What other steps, if any, would you support to limit the amount of metal discarded into MSW?

LEAD

Why is lead pollution of concern? Lead poisoning of children has been described as the most consequential environmental health problem in the country. Of most concern is lead's adverse effect on the nervous system of the developing fetus and small child; thus exposure of pregnant women prompts concern. Although very low doses cause no obvious adverse health effects on the fetus or small child, those low doses are associated with a lowered intelligence quotient (IQ). The deficit may be only several points, but as one author noted, "On a societal basis, the aggregate loss of cognitive acuity due to lead exposure can be enormous." For children who already have a low IQ, the additional deficit could shift them into a severe deficit range. Higher than average body burdens of lead are also associated with increased distractibility and aggressiveness in children. A 1996 *Journal of the American Medical Association* article indicated that boys whose bones contained high levels of lead were more likely to be juvenile delinquents.

To protect her fetus, a pregnant woman needs to limit her exposure to lead. However, her earlier exposure can also pose a problem because the body treats ingested lead in much the same way as it does calcium. Like calcium, lead is deposited in bones, and about 90% of one's lead intake is eventually stored there. During pregnancy, the mother's bones release calcium to the blood to meet needs of the fetus. Lead released at the same time also passes into the placenta and exposes the fetus. Adults are not as sensitive to lead exposures as are children but do suffer adverse effects, including damage to the nervous system and kidney and a greater likelihood of high blood pressure. Other potential adverse effects are anemia and infertility.

What are the sources of lead and routes of exposure to it? Lead-emitting sources in the United States today are similar to those already enumerated for metals in general; they include metal mining and smelting, coal-burning electric power plants, and MSW combustors. *But it is older sources of lead that provide the major route of exposure.* For small children living in old homes containing lead paint, that is inhalation of household dust. The concentration of lead in household dust can be 500 times greater than the background level found in the earth's crust. Lead enters household dust when leaded paint flakes and crumbles into particles small enough to become airborne. Another source of lead that reaches household dust is lead-contaminated soil around older homes, which is tracked inside. In addition, toddlers have a tendency to suck on sweet-tasting leaded paint on window sills. This yields significant exposure in some cases. Lead exposure is a special problem in urban areas because inner-city housing is typically older and thus contains leaded paint. Moreover, this paint is more likely to be flaking than paint in well-maintained suburban homes. Furthermore, inner-city children may have less calcium and iron in their diet than do more affluent suburban children and therefore absorb more of the lead they ingest. Old housing is also more likely to have water pipes that contain lead, and, if not controlled, drinking water can be a significant route of exposure. For people living in lead-free or lead-safe homes, homes in which the lead has been mitigated, exposure through food represents a higher percentage of exposure. Especially under acidic conditions, plants absorb lead from contaminated soil; people are exposed when they eat the contaminated plants themselves or the meat of the animals that fed on them.

Sometimes adults have high occupational exposure to lead. Whereas in America work places are controlled to reduce lead exposure, those in many poor countries are not; for example, small cottage industries in Mexico and Peru – with no controls on exposure – recycle lead batteries from motor vehicles or add lead glazing to pottery. Lead-mining and lead-processing operations are also largely uncontrolled in poor countries. Adverse effects that are observed include high blood pressure and kidney failure.

How are lead emissions controlled and reduced? Because lead is a criteria air pollutant, EPA has established an ambient air quality standard for this metal, and a number of air-emitting sources are already controlled. New EPA regulations will further reduce lead emissions from MSW and medical waste combustors as much as 99%. Lead represents an instance where P^2 has been aggressively and successfully pursued. The P^2 approach that showed the most dramatic effect was banning lead from U.S. gasoline in the mid-1970s; until that time, motor vehicle exhaust was the major source of lead to the environment and the major source of human exposure. Inhaling lead from motor vehicle exhaust remains a major source of exposure in countries that still use leaded gasoline. After the ban in U.S. gasoline, lead emissions to the air fell more than 90%; American blood levels also dramatically declined: In 1995, 9% of American 10-year olds had blood lead levels greater than 10 μg/100 ml, which is the threshold of concern. Compare this figure to that in 1980, when 88% of American 10-year-olds had levels greater than 10 μg/100 ml. For Americans generally, the average blood lead level in 1976 was 14.6 μg/100 ml. By the mid-1990s, the level in children ages 1 to 5 averaged 4–6 μg/100 ml, and only 4.3% tested above10 μg/100 ml. If a child's level is only slightly above 10 µg/100 ml, recommended follow-up steps are primarily finding and reducing sources of exposure to lead; as levels begin to increase, more aggressive attempts are made to find and reduce lead exposures. In some instances, medical intervention is necessary.

Another important P² measure of the 1970s was banning the addition of lead to paint used for domestic purposes. Other P² regulations eliminated lead use in drinking water pipes, in solder used on those pipes, in solder used on food and soft drink cans (although some imported cans still have lead solder), and in drinking water fountains. Ceramics made in the United States are fired at high enough temperature to prevent lead from subsequently leaching into food. Lead print is no longer used by the printing industry. Lead foil, used until very recently to seal wine bottles, has been replaced with other materials. Many states require that lead-acid batteries from motor vehicles be removed from MSW before it is combusted, but unfortunately there is as yet no good alternative battery available.

How is exposure to old sources of lead being reduced? Because lead is already pervasive in the environment, not all P² measures have an immediate effect, so it is important to reduce exposure to old sources. Unlike quickly used gasoline, for example, lead paint remains in place for many years. Virtually any house built before 1980 contains some lead paint and older houses contain much more. However, there are control measures to prevent it from flaking or it can be covered with wall paper or other protective materials. Away from a child's home, day care centers are regulated to provide a lead-safe environment. Municipalities also have screening programs to test blood levels of high-risk children between 6 months and 5 years old. An example of a child at high risk is one living in a home with deteriorated lead paint. Lead in water pipes also remains in place for many years but, again, control measures exist. EPA requires municipalities to maintain drinking water supplies at a pH alkaline enough to limit the corrosion of lead-containing pipes. Individuals can further protect themselves by allowing water from each pipe to run a minute or two in the morning to wash out lead that has leached into the water overnight. If water pipes contain lead, it is also important not to use hot water for cooking because hot water can leach more lead from pipes than cold water.

Remaining lead issues. Despite impressive declines in lead contamination, issues remain. One is that even the much lowered blood level of 4 to 6 µg/100 ml found in children is still about 300 times greater than the estimated 0.016 µg/100 ml of prehistoric people. Another source of concern is that, although average societal levels have greatly declined, the fall has not been uniform.

Among black children living in inner-city housing, 35% still have levels higher than 10 µg/100 ml, as compared to 4% for the population overall. This is because these children live in old, often substandard housing, with deteriorating lead paint.

There is probably no environmental issue that is uncontroversial. Lead exposure is no exception. A survey of high-risk children in Washington state indicated that less than 3% had blood lead levels above 10 µg/100 ml and no child tested had a level higher than 20 µg/100 ml. Thus, the Washington State Academy of the American Academy of Pediatrics, with support from the state Health Department, urged EPA to exempt the state from television advertisements urging that all children be tested for lead. They also argued that, whereas there is little doubt that children with blood lead levels above 30 µg/100 ml suffer adverse consequences, it is not clear that those with levels below 20 µg/100 ml do. They further noted that the lead-screening method now used is considered unreliable. The arguments made in Washington may well be valid; however, it is important to remember that midwestern and eastern cities have much more older housing on average and more poor inner-city neighborhoods, making widespread screening much more justifiable. Nonetheless, lead is not dangerous under all circumstances (review Box 3.4).

At least in the United States and other developed countries, human exposure to lead has been greatly reduced and, with continuing preventive measures, may be reduced yet further. However, lead cannot be completely eliminated from a person's environment. Small amounts are always naturally present in soil, gasoline, paint, or pipes. It is also worthwhile to note that lead was used in consumer products for a reason. In gasoline, it was added as an antiknock agent, a necessary component. When lead was banned, benzene – a human carcinogen – was used instead, although the amount of benzene has also been reduced. Lead was added to paint to protect surfaces against weathering and wear. Thus, some exterior paints deemed to need lead's protective qualities still contain added lead. However, here, too, substitutes are being sought. The product that still uses the most lead, about 63% of the lead processed worldwide, is the lead-acid battery used in motor vehicles, a use that continues to increase (see Table 9.1). Lead use in consumer electronics is also slowly growing.

DISCUSSION QUESTION

How much does it matter to you whether a house that you buy does or does not have lead paint? Explain.

TABLE 9.1 Lead, mercury, and cadmium in MSW (in tons)

YEAR	1986	2000
Lead	213,652	281,887
Lead-acid batteries	138,000	182,000
Consumer electronics	59,000	85,000
Mercury	709	173
Household batteries	621	99
Electric lighting	27	41
Cadmium	1,788	2,684
Household batteries	930	2,035
Plastics	502	380

Note: The two largest sources of the three metals are shown in each case.

MERCURY

Why is mercury pollution of concern? Most people are familiar with the mercury in thermometers and know that it is a liquid metal, sometimes called quicksilver. Elemental mercury can combine with other elements to form either inorganic or organomercury compounds, such as methylmercury. In some locations, mercury levels are increasing. For a pollutant that cannot degrade, an increasing level signals potential problems. In Minnesota, in lakes isolated from industrial activity, mercury contamination of sediments has tripled in the past 150 years; because levels are increasing, it is likely that anthropogenic sources are responsible. Mercury most likely reaches isolated lakes via air transport. In Maine, high mercury levels have been found in an eagle population that shows poor reproductive success. More generally, increased mercury levels in freshwater lakes and methylmercury accumulation in fish are major concerns. Mercury, like lead, provokes most concern because it is toxic to the nervous system and the fetus and small child are the ones most sensitive to its adverse effects. Medical researchers express the concern that even very low levels may delay development. High adult exposure to mercury historically most often occurred in the work place and resulted in a number of toxic effects, including blindness, deafness, and kidney damage. The best known is exemplified by the Mad Hatter in Lewis Carroll's *Alice in Wonderland*, who suffered personality and nervous system problems as a result of mercury inhalation.

What are the sources of mercury and routes of exposure to it? More than is the case with other metals, it is sometimes difficult to determine where the mercury in a given location originated. This is because there are large natural sources in addition to anthropogenic sources. The fact that mercury is volatile poses a second difficulty because it is sometimes difficult to know whether it is newly emitted or is being reemitted. Plants, for example, take up mercury from the air and release it again. A 1993 EPA report indicated that human activities in the United States generate 340 tons of airborne mercury each year, of which combustion generates about 83%. Among combustion sources, fossil-fuel burning power plants are the largest single source in the United States, emitting about 20% of the total. However, MSW and medical waste combustors and commercial and industrial boilers emit mercury, and residential combustion also contributes a few tons. EPA reported that up to 40% of mercury emissions are from natural sources, volcanic eruptions and degassing of the earth's crust and oceans. Some believe that some of this 40% is actually anthropogenic – that earlier emissions for which humans were responsible settled onto land and, after reevaporation, were counted as natural emissions. Worldwide, natural sources generate between 2,700 and 6,000 tons a year of mercury and human activities another 2,000 to 3,000 tons.

Mercury emitted into the atmosphere is

   washed out into water bodies. There

        bacteria transform a portion to methylmercury, which is

             taken up by aquatic plants and invertebrate animals.

             These are *eaten by vertebrate animals*. At successively higher levels on the food chain, methylmercury builds up to progressively higher levels in animal tissues.

FIGURE 9.2. Route of exposure to mercury.

About 95% of human exposure results from eating mercury-contaminated fish (see Figure 9.2). An especially serious mercury poisoning event resulted from consumption of mercury-contaminated fish in Japan. Starting in the early 1930s, and continuing until the late 1960s, Chisso Corporation of Tokyo discharged mercury into Minamata Bay. In the bay, bacteria converted the mercury to methylmercury, which concentrated in fish subsequently eaten by humans. Methylmercury is much more toxic than elemental mercury; furthermore, unlike inorganic forms of mercury, which are poorly absorbed across the human gastrointestinal tract, about 90% of methylmercury is absorbed. Over the years, as many as 200,000 people were adversely affected: Thousands suffered chronic disease and hundreds died from eating contaminated fish. The adverse effects seen included chronic, sometimes severe and debilitating nervous system damage, miscarriages, and deformed fetuses. The Minamata tragedy was at least partially preventable. Authorities were charged, apparently with justification, of knowing about the mercury discharges and their health hazards for some years but failing to stop them or to inform the public. This resulted in a bitter lawsuit between the victims and Chisso Corporation that continued for nearly 30 years. In June 1996, 3,171 victims finally settled. One embittered group, although now ill and elderly, refuses to settle until the government officially apologizes to them.

Most Minamata Bay fish had methylmercury levels ranging from 9 to 24 ppm, but some were as high as 40 ppm. Compare these values to United States FDA guidelines on mercury, which consider fish with less than 0.5 ppm of mercury to be safe for human consumption in any amount. If fish contain 1 ppm of methylmercury, an advisory notifies people to limit consumption; at above 1.5 ppm, people are advised not to eat the fish at all. One part per million of methylmercury is a concentration 10 times lower than that which produced adverse effects in adults in Minamata Bay or in other mercury poisoning events. The FDA reported in 1991 that mercury level in most American fish ranges from 0.01 to 0.5 ppm. The exceptions are large ocean fish, shark and swordfish, which may naturally accumulate up to 3 ppm or more. People eating swordfish and shark are advised to limit consumption to about 7 ounces a week. Individuals who eat large quantities of fish, such as certain Native Americans, are advised to pay special attention to fish advisories. So are pregnant or breast-feeding women or those who may become pregnant. Since FDA's 1991 report, 37 states (some of which have advisory levels lower than FDA's 1 ppm) have issued health advisories on mercury in fish. Some advisories may have resulted from new awareness rather than from increasing levels of mercury. Also, fish were tested from lakes never previously tested. When high mercury levels are found in water bodies, it is not always clear what the source is. An example is the Florida Everglades, where the mercury level in fish is often higher than 1.5 ppm. This condition has led to advisories not to eat the fish, but it is not yet known whether the mercury is anthropogenic or natural. In 1995, mercury contamination was responsible for about two-thirds of all the fish advisories issued nation wide.

Fish is the main route of exposure for most people. However, some individuals, including dental workers and those producing mercury-measuring devices, receive their greatest exposure in the work place. In the late twentieth century, the American work site is controlled to reduce mercury exposure, but, as with lead, workers in poor countries often remain unprotected. Visitors to China describe toxic reactions among those working with mercury as routine and accepted occurrences. Workers are allowed perhaps 2 weeks off work to recover partially in a special home set aside for them. Korean workers have similar high exposures.

How are mercury emissions controlled and reduced? The amount of mercury in drinking water is controlled by EPA through a primary drinking water standard; mercury in food is regulated by the United States FDA and state health agencies by issuing consumption advisories. Mercury is a HAP, and its air emissions are controlled by EPA through the 1990 CAA amendments; new EPA regulations are expected to reduce mercury emissions from MSW and medical waste incinerators as much as 99%. However, the CAA amendments do not cover mercury emissions from power plants, although several states have set stringent limits on emissions. In 1986, 621 tons of mercury was used in batteries in the United States. The amount has been steadily declining and, by the year 2000, this figure is expected to be down to 99 tons (see Table 9.1).

BOX 9.3

Mercury amalgams in teeth. Some human exposure to mercury occurs from the minute amounts of mercury vapor emitted by the amalgams used to fill dental cavities. Critics of this practice claim that the amalgam is linked to multiple sclerosis, arthritis, mental disorders, and other diseases. People with amalgam fillings do have higher mercury concentrations in their blood than do those without them; however, a 1994 article in a dentistry journal reported that although dentists have a much higher body load of mercury than the average American, they have no higher incidence of any disease. The position of the U.S. Public Health Service is that amalgam use should not be regulated unless it is more definitely linked to illness. However, amalgam fillings remain controversial and some dentists no longer use them.

The only batteries that now have mercury added are button batteries and rechargeable batteries. At least one state mandates the removal of button batteries (up to 40% by weight mercury) from MSW before it can be combusted. Another use of mercury is in fluorescent lighting. EPA's voluntary Green Lights program has among its goals the elimination of mercury in lighting and the recycling of spent fluorescent tubes that do contain mercury. Several states control fluorescent lights as hazardous waste because of their mercury content. Despite these actions, mercury use in electric lighting is still slowly growing from 27 tons in 1986 to an anticipated 41 tons in 2000. Mercury was previously used as a fungicide in paint, but after a child's death that resulted from mercury emissions from paint into indoor air, this use was forbidden.

DISCUSSION QUESTIONS

1. The lowest concentration of methylmercury in human blood that has been associated with any adverse effect in adults is 200 ppb. The average concentration of mercury in American blood is about 8 ppb.
   (a) Do you believe 8 ppb gives you a comfortable margin of safety? Explain.
   (b) Assume a group of adults, including women of childbearing age, have 25 ppb methylmercury in their blood. No one in this group suffers adverse effects. Would you feel confident that a fetus would likewise suffer no adverse effect from this level? Explain.
2. In 1996, the Department of Defense (DOD) proposed to sell the U.S. stockpile of mercury (60% of the world's supply) on the open market.
   (a) From an environmental viewpoint, do you believe that this is a sound idea? Explain.
   (b) In Brazil, mercury is used in gold mining, a process that has, since 1989, released about 170 tons of mercury into the environment. Brazil imports mercury from other countries. Thus, if the DOD sold mercury, Brazil could buy part of it. Does knowing this affect your viewpoint?
   (c) Most nations now exporting mercury to Brazil restrict the use of mercury within their own borders. Should they be selling mercury to Brazil? Explain.

CADMIUM

Why is cadmium pollution of concern? Whereas lead has been mined for at least 4,000 years, cadmium was not discovered until 1817 and has been heavily mined only since the end of World War II. Environmental cadmium levels are lower than those of lead and people ordinarily have lower exposure. Nonetheless, exposure levels are hundreds of times greater than they were among preindustrial people and this metal generates growing concern. Cadmium bioaccumulates in the kidney, and the amount of cadmium stored in this organ increases with age. In some individuals, it concentrates to levels not much lower than those known to damage the kidney. For those with high occupational exposure to cadmium, kidney damage has been the most prevalent chronic effect seen. Because cadmium affects calcium metabolism, bone degeneration has also been observed. Cadmium can also accumulate to high levels in liver. In laboratory studies with rodents, depending on concentration, cadmium shows numerous toxic effects, including birth defects, and is a carcinogen.

The most notorious human poisoning with cadmium occurred among poor elderly Japanese women who experienced *itai itai* disease, characterized by kidney damage and by severe bone damage that left their bones brittle and painful. This second effect led to the name *itai itai* (pain-pain) disease. The rice these women ate was grown in paddies near smelting operations, and the paddy soil contained cadmium levels up to 10 times those found in other soils. Because their diet consisted primarily of rice grown in these paddies, the women had higher cadmium exposure than those who ate less rice and had a more varied diet. For example, a diet high in calcium, iron, and fiber is associated with lower blood levels of cadmium. This example demonstrates the need for a varied diet, which is not always available to the poor. Most of the affected Japanese women had borne several children, perhaps leaving their bones especially susceptible.

What are the sources of cadmium and route of exposure to it? As with lead and a number of other metals, cadmium is released into the air by mining and smelting operations and by fossil fuel combustion, especially of coal. Polyphosphate fertilizers and sewage sludge have cadmium concentrations higher than those normally found in soil and contribute to cadmium buildup in agricultural soils. Cadmium is best known to consumers from its use in the nickel-cadmium rechargeable batteries used to power small appliances, so-called Nicad batteries, which are a major source of cadmium in MSW.

As is often the case with pollutants, the route of exposure can be quite different from its sources. More than 90% of the average nonsmoker's exposure is through food. Shellfish concentrate cadmium, and consumption of scallops and oysters can be a major exposure source for people who eat them. Fish concentrate cadmium to a lesser extent and are a lower source of exposure. Liver and kidney concentrate cadmium and contribute to the exposure of those who consume these organs. Advisories have been issued to hunters in some locales warning them not to eat liver and kidney in moose or bear. *High cadmium levels in soil present special concerns because plants take up cadmium more readily than other metals.* Because tobacco plants concentrate cadmium, smoking a pack of cigarettes a day can double a person's exposure. Worse, an adult absorbs more than 90% of the cadmium inhaled through smoking, as compared to only about 5% of the cadmium ingested in food. The greatest route of exposure for workers with occupational exposure is inhalation.

How are cadmium emissions controlled and reduced? There are a number of controls on cadmium. EPA has a drinking water standard, which limits exposure to cadmium by this route. Its air emissions are controlled by so-called best available technology (BAT), according to the mandates of the 1990 CAA amendments. As with lead and mercury, new EPA regulations will reduce cadmium emissions from MSW and medical waste incinerators by as much as 99%. However, emissions from power plants are not controlled. Two current major uses of cadmium, noted in Table 9.1, are in batteries and plastics. Although cadmium is no longer added to disposable batteries, it is used in Nicad rechargeable batteries. As with disposable batteries, the final fate of rechargeable batteries is to be discarded. Thus, Nicad batteries continue to be the greatest source of cadmium in the solid waste stream, and, as Table 9.1 shows, the amount continues to grow.

DISCUSSION QUESTIONS

1. As an environmental pollutant, what factors can sometimes make cadmium as an environmental pollutant an even greater concern than lead or mercury? Explain.
2. Individuals in industrialized countries have a cadmium intake ranging between 10 to 50 micrograms per day. EPA recommends 70 micrograms per day as an upper limit.
   (a) Do you believe that the current consumption rate provides a comfortable margin of safety? Explain.
   (b) Now, remember that use of cadmium is increasing. What further actions, if any, should society be taking to limit cadmium releases to the environment?
   (c) What can you as an individual do?

ARSENIC

Arsenic was deliberately used for hundreds of years as a poison. A person fed small doses of arsenic would finally die of what appeared to be pneumonia. Probably because arsenic is easily detected by the modern forensic scientist, it is now seldom used as a poison. A more positive use of arsenic was as a syphilis treatment until it was displaced by modern antibiotics. Chronic arsenic inges-tion is associated with skin cancer and chronic arsenic inhalation with lung cancer. Western desert soils are naturally high in arsenic. Some waters contain naturally high levels of arsenic and pose a chronic risk to those drinking it. Arsenic is sometimes referred to as a heavy metal, but it is not. In fact, it is not a metal at all, strictly speaking, but rather a "semi-metal" or metalloid; how-ever, metal-processing operations and fossil fuel combustion are significant sources of arsenic. Also, like a number of metals, arsenic continues to build up in the environment. Before the advent of synthetic organic pesticides, arsenic was used as a weed killer and rat poison. Along with chromium and copper, it is still a wood preservative in pressure-treated wood. Except in occupa-tional settings, most human exposure is from food. Fortunately much of the arsenic in food may not be bioavailable. There is a drinking water stan-dard for arsenic, and it is a HAP whose air emissions are controlled under the 1990 CAA.

SELENIUM

Many other metals could be discussed, each raising special issues. Selenium is especially interesting; it is an essential nutrient, and most of our exposure to it is through food. Dietary selenium protects against the toxicity of several met-als, including lead, cadmium, and arsenic. It may also act as an anticarcino-gen. Epidemiological studies indicate that people living in locales with high selenium levels in the soil have lower rates of several cancers.

Selenium does not always present such a friendly face. Although it is natu-rally present at quite high levels in western soils, agricultural crops can safely be grown there. But irrigation is necessary in California. Excess irrigation wa-ter, containing selenium dissolved from the soil, is drained back to ponds. There, as water evaporates, selenium concentrates to levels high enough to endanger water fowl and other wildlife. Along the return to ponds, the irriga-tion waters have also contaminated groundwater with selenium. Societal con-trols on selenium exposure include a drinking water standard, and because it is a HAP, its air emissions are regulated by the 1990 CAA. These measures, however, are not of a nature that could help control high selenium levels in irrigation return water.

TRENDS IN METAL POLLUTION

Metals continue to be heavily used in consumer products and in industrial and military equipment. This means continuing environmental damage from mining and metal processing. Developed countries such as the United States are working to deal with this damage, and ongoing research and development may lead to the use of lighter nonmetallic materials in automobiles, aircraft, and consumer products. In the meantime, developed countries are practicing control measures to limit metal emissions. Unfortunately, emissions remain very high in many poor countries. Aggravating the situation, mining and metal-processing industries sometimes move from the developed nations to countries with less strict environmental standards. Even so, multinational corporations have a better environmental record than do state-owned industries in poor nations. Other trends are as follows:

- Many countries, including the United States, burn coal to generate electric power and may do so for many more years. Metal emissions from coal burning will probably be controlled in the United States or clean coal technology research expanded to find means to remove metals as well as sulfur. If these steps are not taken, metal accumulation in the environment from burning coal will continue to increase. Two very large poor nations, China and India, are burning more coal as part of their development plans and doing so with no or minimal controls.

- The United States has major problems with motor vehicle emissions, including the metal emissions in their exhausts, but there are ongoing efforts to reduce these. At the same time, the number of motor vehicles in many poor nations is rapidly growing, and metal emissions from this source are likewise increasing. Unlike the United States and a number of other developed countries, many poor nations still burn leaded gasoline.

A positive note was sounded in a 1995 article in *Science* written by Carroll Hodges of the United States Geological Survey who noted that the demand for accountability from mineral mining and processing industries is rising worldwide and that environmental protection and rehabilitation are becoming high priorities. Twenty-seven national and international metal companies recently formed a council that has the goal of promoting sound environmental and health practices to ensure that metals are safely produced, used, recycled, and discarded. One mining company operating in Indonesia announced that it will make investments to ensure that the mine operates under the same environmental standards used in the United States.

For thousands of years, basic mining techniques have remained largely unchanged. Crude ores were dug from the earth, were crushed, and the minerals obtained by high-temperature smelting or roasting of the ores or by extraction of the ores using highly toxic chemicals like cyanide. For example, phosphates are extracted from their ores by burning them at high temperature or by treating the ores with sulfuric acid. Both techniques are hazardous and produce large amounts of waste. But now there are hopeful signs that P^2

will become a part of the picture. This will mean reducing hazardous chemical use in mining, roasting, and smelting; reducing emissions; and reducing
energy use. Two approaches follow:

- It has been known for many years that some bacteria can use minerals as a
  source of the energy they need to grow. The best known of these is *Thio-
  bacillus ferrooxidans*, a microbe capable of growing on low-grade copper
  ores. As the bacteria grow on the ore, copper enters into solution and can
  be recovered. A similar approach has been developed in phosphate mining. Some hope this process, referred to as *biomining*, may eventually be
  able to recover gold from even low-grade ores. The promise is so great,
  both for P^2 and for obtaining of minerals at lower cost, that intensive research and development efforts are being pursued to overcome biomining's significant problems. Recall that microorganisms are already used to
  treat hazardous waste. The use of microbes in biomining, however, reaches
  to the top of the waste management hierarchy to P^2.
- Another promising P^2 approach is in situ mining being explored by the
  United States Bureau of Mines in cooperation with two mining companies.
  In this technique, a weak chemical solution is injected underground to put
  minerals in rock into solution without disturbing the surface of the land.
  This technique holds the promise of generating almost no waste or other
  environmental health and safety problems.

FURTHER READING

Bradbard, L. 1993. Dental Amalgam, Filling a Need or Foiling Health? *FDA Consumer*,
  27(10), 22–25, Dec.
Foulke, J. 1994. Mercury in Fish. *FDA Consumer*, 28(7), 5–8, Sept.
Hodges, C. A. 1995. Mineral Resources, Environmental Issues, and Land Use. *Science*, 268,
  1305–12, June 2.
Mercury: An Atmospheric Hitchhiker. 1990. *Health & Environment Digest*, 4(4), 1–4, May.
Miller, R. W. 1988–1989. The Metal in Our Mettle. *FDA Consumer*, 22(10), 24–27, Dec./Jan.
Nriagu, J. 1990. Global Metal Pollution, Poisoning the Biosphere? *Environment*, 32(7),
  7–11, 28–33, Sept.
Nriagu, J. 1996. A History of Global Metal Pollution. *Science*, 272, 223–4, April 12.
Renner, R. 1995. When Is Lead a Health Risk? *Environmental Science & Technology*, 29(6),
  256A-61A, June.
Sharma, A. A. 1996. Metal Mining: How Does It Pollute Water? *On Tap*, 5(1) Spring, 16–17.
Waalkes, M. P. 1991. Cadmium and Human Health. *Health & Environment Digest*, 5(4), 1–3,
  May.

10

PESTICIDES

INTRODUCTION

Pesticides are chemicals or other agents used to destroy any organism that is considered a pest. Almost any living creature can be a pest in certain circumstances. Pesticides are used to increase the production of food and fiber and to promote public health. A pesticide that kills insects is called an insecticide and one that kills plants an herbicide; other examples of pesticides are given in Table 10.1. Among the pests attacking agricultural crops are insects, fungi, rodents, and birds. Weeds, loosely defined as any undesired plants, are also agricultural pests. The amount of manual labor necessary to control weeds without help from herbicides or machines may be difficult for many Americans to imagine; an African woman in the United States to study pest control made a comment at a 1992 conference that rural women in her country spend 60% of their working day just pulling weeds. The United States has such an abundant food supply that it may also be difficult to imagine the total devastation of a crop. A famous example of such devastation is the nineteenth-century fungal infection of potatoes in Ireland. The famine it produced resulted in the immigration of millions to the United States. Pest infestations have been problems to humans for as many thousands of years as humans have practiced agriculture. For longer periods yet, pests, including fleas, lice, mosquitoes, flies, roundworms, rats, and mice, have threatened human health.

For thousands of years, people looked for means to rid their crops of the insects eating them, the weeds choking them, or the fungi making them uneatable. People began using sulfur, a chemical still used by some organic gardeners, as a pesticide thousands of years ago. Extracts of chrysanthemum flowers containing pyrethrum have been used for nearly as long, and tobacco extracts containing nicotine have been used for hundreds of years. Starting in the late 1800s, chemical pesticides containing arsenic, mercury, lead, and copper came into widespread use. An elderly man wrote a letter to a periodical in

TABLE 10.1   Some categories of pesticides

PESTICIDE [a]	PEST KILLED
Insecticides	Insects
Larvicides	Insect larvae
Fungicides	Fungi that grow on plants and, sometimes, animals
Herbicides	Plants (weeds)
Fumigants	Many life forms
Disinfectants	Microorganisms found on surfaces

[a] There are also pesticides specific to mites; algae; birds; snails, slugs, and other mollusks; nematodes; fish; and other organisms.

1989 describing his grandmother's 1920s gardening chemicals; in addition to her occasional use of the highly toxic gas hydrogen cyanide as a fumigant, she used paris green (copper arsenate), lead arsenate, and nicotine sulfate to control garden pests. Given the widespread use of metal pesticides in the first half of this century, it was not surprising that the first household hazardous waste roundup that Massachusetts carried out in the 1980s, recovered 3 tons of arsenic in chemicals that had been sitting in sheds and barns for many years.

Even in large amounts, sulfur and copper only partially controlled pests. So it is not surprising that when the very effective synthetic insecticide dichloro-diphenyltrichloroethane (DDT) was introduced in 1942, it was quickly embraced. DDT was lethal to many insects. It killed the mosquitoes and flies that spread disease, the insects infesting crops, and other insects, such as body lice. It was considered a tremendous contribution to public health, and the discoverer of its insecticidal activity received a Nobel Prize in medicine in 1948. Many other synthetic chemical pesticides were quickly developed and saw widespread use. Even in the 1940s, the ability of insects to mutate and become resistant to pesticides was observed; however, most pesticides remained widely effective and the phenomenon of resistance caused little concern. DDT and similar organochlorine pesticides showed relatively low acute toxicity to humans and were not absorbed through the skin. Possible chronic toxicity was little considered. The result was wide and often indiscriminate use of pesticides. It was not until the early 1960s that Rachel Carson's famous book *Silent Spring* forced Americans to see the darker face of DDT and other pesticides.

Why are pesticides used?

The use of pesticides makes it possible to grow crops at times and in places where they could not otherwise be grown. Fruits and vegetables are on the market year round not only because they can be transported long distances from warmer climates, but because pesticides make it possible to grow them over longer growing seasons and in a greater number of locations. Without fungicides, for example, certain crops could not grow in locales or in seasons when fungi grow prolifically. Many believe that the health advantages of fresh fruit and vegetable availability year round and their lowered cost make up for any human health risk posed by pesticides. Another public health benefit is reduced growth of fungi on treated crops, fungi which can produce very toxic chemicals. Pesticides make monoculture possible; that is, large tracts of land can be devoted to only one crop, for example, wheat, cotton, soybeans, or corn, season after season at one location. Without chemicals, the pests that attack a monoculture crop would build up until the crop could no longer be grown at that location. Pesticides also make it possible to store food products for long periods. After harvest, grain is fumigated to kill the insects and disease-causing organisms infesting it. These organisms could otherwise multiply during storage, destroying part or all of the grain. For similar reasons crops are fumigated before being transported long distances to market. Pesticides are also used to control the vectors that spread disease, such as mosquitoes, flies, ticks, and rats.

Disinfectants ("germicides") are used to kill microorganisms that live outside the body. Regulated as pesticides by EPA, disinfectants have been used since 1867, when Lister began using phenol to disinfect operating rooms. Chemicals related to phenol are still widely used disinfectants. The active chemical in the commercial product Lysol is an example. Other common disinfectants used in home and industry are chlorine-containing compounds such as sodium hypochlorite (common household bleach). The antibiotics used to kill microorganisms in humans and animals are regulated by the U.S. FDA, not by EPA.

DISCUSSION QUESTIONS

1. Think about the reasons that pesticides are used.
   (a) Which do you believe are good reasons?
   (b) Do you disapprove of some of the ways that pesticides are used? Explain.
2. (a) On what occasions do you believe it is appropriate to use a pesticide in your own home, yard, or garden? Explain.

(b) Are there home uses that you consider trivial?

(c) Recall, if you can, any recent TV advertisement for pesticides that you have seen and your reaction to the advertisement.

Major insecticide categories

There are many types of pests, and, not surprisingly, many types of pesticides have been developed to fight them. Because it is not possible to describe them all, only some well-known insecticides will be discussed.

- DDT is the best known of the polychlorinated insecticides. Lindane, aldrin, and heptachlor are other polychlorinated insecticides, also widely used for years. Like DDT, most polychlorinated pesticides have been banned or their use greatly restricted in this country because of their persistence in the environment, damage to animal populations, and ability to bioaccumulate in animal fat. Both microorganisms and animals find them difficult to degrade. Direct sunlight can destroy them, but they are often trapped in sediments or other locations without sun exposure. Over the 30 years in which it was used, DDT accumulated in many locations, including the Great Lakes region. With time polychlorinated chemicals do break down, and, 20 years after its banning, DDT's environmental levels had dramatically declined in the Great Lakes area, as had its level in fish and in the blood, breast milk, and tissues of humans eating the fish. Nonetheless, DDT still is detectable in the United States and will be for years to come.

     Polychlorinated chemicals that reach the Arctic pose a special concern because there extreme cold prevents them from being degraded as they are in warmer locations. Other defining characteristics of organochlorine pesticides are their low solubility in water and high solubility in lipid materials, including animal fat, in which they bioaccumulate. Fortunately, because of their low water solubility, they cling to soil particles and have little tendency to dissolve in rainwater and migrate into water bodies.

- As highly chlorinated pesticides were being banned in the United States, organophosphate insecticides became more widely used. These have much shorter lives in the environment and do not bioaccumulate to high levels in fat. Their disadvantage is that they are more acutely toxic than organochlorine pesticides, sometimes much more toxic. Whereas DDT, for example, has an LD_{50} of 113 mg/kg in rats, the LD_{50} of the organophosphate insecticide parathion is 3.6 and 13.0 mg/kg in male and female rats, respectively. Some organophosphate pesticides – again parathion is an example – can also be absorbed across the skin, greatly increasing the danger to pesticide applicators. Most acute human pesticide poisonings are caused by organophosphate pesticides, which are chemically related to the exceedingly toxic organophosphate nerve gases. Another concern centering around organophosphate use is their water solubility, which allows them to run off into water bodies or percolate down into groundwater.

- Although a few are exceptionally toxic, most members of the carbamate family of insecticides are less acutely toxic than the organophosphate pesticides. Like organophosphate pesticides, they are short-lived in the environment. Because of their lower toxicity, they are often found in products used by homeowners but are less useful to farmers.

How pesticides exert their effects

- All three of the insecticide groups just discussed are neurotoxic (toxic to the nervous system). The organochlorine pesticide DDT acts on nerve membranes to prevent normal conduction of nervous impulses. Organophosphate insecticides inhibit the action of the enzyme that breaks down the neurotransmitter acetylcholine; as acetylcholine levels build up, the result is uncontrolled firing of nerves. Carbamate insecticides exert toxicity in a similar manner, but their toxic effects are shorter-lived. There are many other ways that insecticides can kill target pests; for example, the botanical insecticide (pesticide derived from plants) rotenone is a stomach and contact poison.
- Herbicides fall into many chemical groups. Some interfere with the normal function of plant cell membranes, others act on plant metabolism to cause abnormal growth, and still others inhibit the action of enzymes necessary to plant life.
- Some pesticides are selective: They act against a limited group of organisms because they affect some aspect of metabolism specific to a limited number of plants, animals, or microbes. Any chemical can be toxic in high enough doses. However, an herbicide that interacts with an enzyme found only in plants is less likely to harm birds, other animals, and humans. Other pesticides are broad-spectrum, affecting a wider range of organisms and more likely to pose a danger to nontarget species. Fumigants are an example; the fumigants hydrogen cyanide and methyl bromide affect biochemical respiration in many species. A fumigant is often deliberately used to kill a variety of pests, those infesting the grain stored in an elevator or a greenhouse, for example. Fumigants are also used to sterilize soil or seeds. They are often gases that can penetrate an enclosed space to do the job required of them.

WHAT ARE THE POLLUTANTS OF CONCERN?

In the same way that almost any living creature can be a pest at times, almost all pesticides can become pollutants because of the way they are applied, most often by aerial spraying. In some cases, contamination may not be a problem; the pesticide may swiftly degrade into innocuous products, it may not be very mobile and thus not reach water, and it may have low toxicity to nontarget organisms. Unfortunately, in most cases, pesticide contamination is a concern. An insecticide is likely to come to mind when the subject of pesticides is discussed, but herbicides account for about half of the total volume of American pesticide use. They are used on most agricultural cropland. Insecticides and fungicides are used in lower amounts and on significantly fewer acres.

WHY ARE PESTICIDES OF CONCERN?

Human exposure

Most pesticides are intended to kill specific pests or groups of pests while having minimal toxic effects on humans and other nontarget species. However, both wildlife and humans can also be at risk. Recall from Chapter 4 that EPA's SAB rated worker exposure to chemicals in industry and agriculture as a high-priority environmental health risk. Even though most are trained in safety techniques, field workers mixing, loading, and applying pesticides are among those at highest risk; they may inhale the pesticides or sometimes absorb them through the skin. Landowners may not properly educate laborers, especially temporary migrant laborers. Even trained workers may become casual and not use protective clothing and equipment properly. Workers entering treated areas too soon after pesticide application can also be at high risk. Others who may be exposed are children playing in or near a treated area. Small amounts of exposure occur through drinking water. Another source of exposure to humans, food consumption, is discussed later.

Information from the World Health Organization indicates that adverse effects of pesticides on human health are much worse in third world countries, as at least 20,000 deaths and a million illnesses every year result from pesticide misuse. Large plantation owners in these countries are often casual about worker exposure and do not provide workers with personal protection or the education needed to use pesticides properly. Furthermore, the pesticides used may be more toxic than those used in the United States, and the language on labels is most often English, which workers typically cannot read.

Consumer exposure to pesticide residues on food

A specific pesticide can be legally applied only to certain crops, which cannot be legally harvested until a specified number of days after application, the length of time varying with the pesticide and crop. However, there have been cases in which farmers inadvertently or deliberately mishandled a pesticide by applying it to a crop for which it was not registered or harvested the crop too soon after application. In 1985, aldicarb – one of the most toxic pesticides in use – was illegally applied to watermelons in several western states and British Columbia. More than 1,000 people became ill after eating the melons. Such dramatic incidents are the exception; a greater concern to many is chronic exposure to trace levels of pesticide residues in food. A tolerance level is set for each pesticide on the crops for which it is used; residues greater than the tolerance level should not appear on the harvested crop. In 1992 and 1993, the United States FDA, which monitors foods for pesticides, looked for 66 different pesticides on 42 fruits and vegetables. Of about 15,000 samples tested, 3.1% violated standards. However, an environmental organization, the Environmental Working Group, charged in 1995 that the violation rate was really 5.6%, a figure that FDA disputes. The group recommended a requirement that the food industry test for pesticides under FDA supervision before their food products could be marketed. This would assure that only foods with pesticide residues below tolerance levels were sold. Such a system would be similar to a new procedure used to reduce microbial contamination of seafood. On the reassuring side, it is important to point out that the produce samples tested by FDA were raw, unpeeled, and unwashed; testing of produce fresh from the field represents a worst case picture because the pesticide residues continue to degrade thereafter on the way to market and up to the time that the food is eaten. Peeling or removing outer leaves usually removes part of the pesticide, and, if the food is cooked, yet more may degrade. On the less reassuring side, FDA's routine testing does not pick up all pesticides registered for use on a given crop although specific tests are available for other pesticides if there is reason to suspect their presence.

Whether pesticide residues are ordinarily present on foods at a level that could adversely affect human health has been an ongoing controversy. In 1993, an NRC committee reported that children were not adequately protected by current methods of examining pesticide toxicity, exposure, and standard setting. To determine a pesticide's RfD, the risk assessment procedure described in Chapter 4 is used. However, unlike EPA's health-based air or water standards, pesticide tolerance levels are not health-based; they are based on standard agricultural practices. The writers of the NRC report expressed their concern that children ingest more pesticides than adults and may be more susceptible to them. On the basis of the report, the EPA, FDA, and

USDA began developing new procedures to examine pesticide effects on development and on the nervous, immune, and endocrine systems. In 1996, EPA administrator Carol Browner announced a new policy: Chemical risk assessment will begin to take into account health risks specific to infants and children. The new approach will apply to chemicals generally, not just to pesticides. In the meantime parents are urged to continue to feed their children large amounts of fruits and vegetables, a recommendation based on the belief that the beneficial effects of fresh foods outweigh small pesticide risks.

There is another side to the argument as to potential harms of pesticide residues. Dr. Bruce Ames of the University of California points out that not only synthetic chemicals and pesticides but a great many natural chemicals in our food are carcinogens or otherwise toxic. For example, molds growing on grains and nuts produce toxins; to date, 16 of these have been found to be animal carcinogens. Roasted coffee contains about 300 chemicals, 11 of which have been tested for ability to cause cancer; 8 of the 11 were carcinogens in mice. One cup of coffee contains about 10 mg of these chemicals, the equivalent of several months' intake of synthetic carcinogens. Plant toxins can have adverse effects; for example, the aflatoxins produced by molds not only are carcinogens, but adversely affect the liver. Potato alkaloids are toxic to the central nervous system; in one instance, a potato was developed that resisted insect infestation. Unfortunately, the chemical alkaloid, produced by the potato in amounts large enough to repel insects, was also toxic to humans. Certain mushrooms produce chemicals well known for their high, sometimes extreme, toxicity. There are many other examples of plant toxins that demonstrate a variety of adverse effects. In fact, poison control centers sometimes publish brochures cautioning against a number of both cultivated and wild toxic plants.

Some believe that plant toxins are less of a concern than synthetic ones because humans evolved in their presence. However, the body's defense mechanisms are general, not specific to toxicants with which animals evolved. Furthermore, human diets have changed drastically and continue to change. Many foods we now eat were not in our diet before the practice of agriculture; many others were added much later. Potatoes have been widely consumed for only about 400 years. Tomatoes became part of the diet only in the early 1800s. Other fruits and vegetables, once only seasonally available, are now consumed year round. Another point made by Dr. Ames is that whereas synthetic pesticides detected in foods are present typically at parts per billion levels, natural toxins are often present at the million-fold higher level of parts per thousand. He estimates that 99.9% of dietary pesticides are natural: An average American consumes 1,500 mg/day of natural pesticides and their breakdown products, whereas the estimated intake of synthetic pesticides is only 0.09 mg/day. Ames and others also observe that synthetic pesticides have markedly lowered the cost of the fruits and vegetables that many believe lower the risk of cancer, heart disease, and other disease.

In 1996, an NRC report emerged on carcinogens and anticarcinogens in food. The report made several points that support those of Dr. Ames. It also pointed out that foods contain anticarcinogens as well as carcinogens, and that the great majority of synthetic and natural chemicals are unlikely to pose an appreciable cancer risk. Professor S. M. Cohen of the University of Nebraska Medical Center has stated "whether a chemical is man made or God made, it is a chemical and the body handles it in similar ways." The report did raise the question of how we can assess the interaction, whether natural or synthetic, of the hundreds of chemicals ingested daily, each in nontoxic quantities. One point stressed was that health risks from chemicals in food pale in comparison to those posed by gluttony, especially the overeating of foods rich in fat.

Other problems associated with pesticide use

There is a viewpoint that sees the pesticide residue controversy as of little consequence relative to the more general issue of modern agricultural practices. With its high input of pesticides, fertilizers, energy, water, and machinery, modern agriculture may not be sustainable in the long term. Some problems associated with ongoing large-scale pesticide use follow.

Effects on nontarget species. Nonhuman species often have much higher pesticide exposures than do humans. Birds, other animals, and beneficial insects may suffer adverse effects or die as a result of pesticide exposure. In California, an estimated 11% of honeybee colonies are lost each year because of pesticide exposure. On the East Coast, the death of as many as 2 million birds in one incident was associated with the use of the insecticide, carbofuran. *Bird poisonings associated with pesticides still regularly occur.* Birds are killed not only by pesticide spraying, but through eating of pesticide-treated seeds. Even in the 1990s fish and other aquatic species die or suffer adverse effects caused by pesticide runoff from land or pesticide spills.

Effects on pest populations. One effect that pesticide use has is on secondary pest population growth, an effect arising in the following way: The primary pest that the farmer wants to kill may ordinarily hold down the population of a potential pest, the *secondary pest*. As the primary pest is killed off, the secondary pest, freed of its enemy, undergoes population growth, until it becomes a pest. Another effect is *pest resurgence.* The insecticide kills off a pest predator as well as the pest. Thus, when survivors of the pesticide reproduce, their population rebounds rapidly because the predator that previously helped to control it is gone.

The most serious problem is the development of *pest resistance.* A few individuals in the target population (whether insects, plants, or other pests) may have a genetic mutation that permits them to tolerate the pesticide. When these individuals reproduce, they pass the genes for resistance to their offspring. Over time, the resistant population increases until few individuals susceptible to the pesticide are left. Sometimes, when they meet pest resistance, applicators respond by applying larger quantities of the pesticide. The pest then become resistant to these higher doses. Or, if the applicator switches to a different pesticide, resistance develops to it as well. Because some pest species reproduce very quickly and in huge numbers, resistance can sometimes develop rapidly. Almost 500 insect species and 150 plant pathogen species resist one or more pesticides, and 113 weed species resist all known chemical pesticides. A serious example of resistance is shown by the *Anopheles* mosquitoes that carry the malaria parasite. Malaria affects hundreds of millions of people worldwide and causes about 1.5 million deaths a year. *Anopheles* mosquitoes were once very susceptible to DDT and other insecticides; now, they are increasingly resistant and progressively larger amounts of pesticide are required to control them. At the same time, the incidence of malaria in some locations is higher than it was before pesticides were introduced. The process of pests developing resistance to pesticides is entirely analogous to microorganisms' developing resistance to antibiotic drugs. In the case of antibiotics, this situation has become so serious that physicians now refer to an approaching post-antibiotic age.

Indirect problems associated with pesticide use. Some problems result indirectly from pesticide applications. For example, the monoculture crops that pesticide use makes possible often cover large tracts of land. These are farmed by using heavy machinery, which leads to soil compaction and the problems associated with compaction. Soil quality is worsened further by the fact that neither pesticides nor synthetic fertilizers provides the organic material to the soil that manure and compost do. Lack of organic matter makes the soil less hospitable to the worms and microorganisms needed for good soil fertility.

Although pesticides are used to reduce pest populations and thus lessen crop losses, about a third of crops are nonetheless lost to pests. This figure is not much different from what it was before the advent of modern pesticides and holds true despite the fact that pesticide use has increased greatly over the years. However, it is argued that the figure we should examine is the actual food yield, which has at least doubled. So although the percentage of food lost is the same, food yield has increased. Nonetheless, as increasing numbers of pest species become resistant, many agricultural scientists doubt that effective and relatively safe new pesticides can be developed fast enough to cope with them. Thus, the question remains: Is the increased food production seen in recent years sustainable?

DISCUSSION QUESTIONS

1. Are the principles of antibiotic use any different from those of pesticide use? Explain.
2. Velpar (hexazinone) is an herbicide used on Maine's blueberry barrens. In 1995, after 10 years of use, Velpar was detected in the groundwater of locales where it is used at an average concentration of 5 ppb. The drinking water of one school had less than 2 ppb, but school board members decided to install a filter to remove it. Some individuals want to forbid blueberry growers to continue to use Velpar. EPA's health-based limit for this herbicide is 210 ppb.
   (a) Knowing this, would you be concerned about 2 ppb in your drinking water?
   (b) Would you feel any different if the 2 ppb were in water drunk by your child?
   (c) Do you believe that pesticides that migrate into groundwater should be prohibited from use or restrictions placed on their use? Explain.
   (d) If current levels of Velpar do not concern you, would you support continued groundwater monitoring? Why?
3. Disregarding potential human health effects, long term-use of Velpar or other herbicides can lead to loss of organic material from soil. So can long-term use of synthetic fertilizers.
   (a) What are possible reasons for loss of soil's organic material under these conditions?
   (b) Why is loss of organic matter a concern?

WHAT ARE THE SOURCES OF PESTICIDE POLLUTION?

The majority of United States pesticide sales are to farmers, but many others also buy pesticides:

- Homeowners use pesticides to kill the hornets in a nest too close to the home; the unwanted dandelions on the lawn; the slugs or insects in the garden; flies, cockroaches, ants, moths, or rodents in the home; mildew in a damp bathroom; or a pet's fleas.
- Utility owners use them to keep rights-of-way clear.
- Golf course owners use pesticides to maintain weed-free greens.

- Commercial establishments use them to keep the premises free of insects, mold, and other pests.
- Industry employs pesticides to control mold and algae growth.
- Governments use them to control rats, insects, or other pests that carry disease or present other dangers.

What may be surprising is how few people do not use pesticides. However, in terms of volume used, agricultural lands are the major source of pesticides in the environment.

Pesticide movement in the environment

Herbicides and insecticides are applied over large areas of agricultural fields and forests, and farmers may apply them a dozen times or more during a growing season. For each application, less than half of the pesticide actually reaches the insect, weed, or other pest. Most become a pollutant. Sometimes foggy weather prevents pesticides from being airborne away from the point of application, posing a problem to those exposed to the trapped pesticides. Most pesticides are applied to crops by spraying. They then drift with air currents from the point of application; the largest amounts settle onto land and water close to the point of application, but smaller amounts, swept higher into the atmosphere with the winds, can be carried thousands of miles. Certain polychlorinated pesticides detected in wilderness lakes in the northern United States or Canada are not used in these countries and are assumed to have been blown from Mexico or other Latin American countries. Once soil and water become contaminated with these persistent pesticides, they may remain so for many years, especially in northern locations, where cold weather and lack of intense sunlight prevent them from degrading.

Remember from Chapter 7 that agricultural lands are a major nonpoint source of pesticides, fertilizers, eroded soil, and manure. Runoff from lands to which pesticides have been applied is responsible for most surface water contamination with pesticides. A comprehensive study of the Mississippi River basin, for example, detected more than 40 pesticides. One, the herbicide atrazine was present in 95% of the samples. In localized areas, atrazine even exceeded EPA's MCL of 3 μg/L. Pesticides are also found in runoff from municipal streets and grounds, areas sprayed by utilities, golf course greens, and yards and gardens of homeowners. Water-soluble pesticides not only can be found in runoff, but can move down through soil to reach groundwater. Chlorinated pesticides are much less water-soluble and cling to soil particles, so they are less likely to contaminate groundwater. An EPA study found that 10% of American community drinking water wells and 4% of rural wells contained detectable – not necessarily significant – amounts of at least one pesticide.

Groundwater contamination is a greater concern than surface water contamination because even pesticides that have only short lives in surface water may degrade very slowly in groundwater. Groundwater is also much harder to clean up. Recall that municipal drinking water, whether from groundwater or surface water, is tested for a variety of pollutants, including a number of pesticides. The water must be remediated if a contaminant is found above its MCL. Because of this practice, significant human exposure is less likely to occur through municipal drinking water systems; however, private well owners are not required to test.

HOW IS PESTICIDE USE CONTROLLED AND REDUCED?

Pesticides are regulated under the Federal Insecticide, Fungicide and Rodenticide Act (FIFRA), a law administered by EPA. FIFRA regulates the active ingredients in pesticide products; it is the active ingredients that actually kill the pest. Unlike the Delaney Clause, discussed in Chapter 1, which mandated zero risk, FIFRA mandates that pesticide risks be balanced with their benefits. Some benefits were discussed earlier in this chapter. FIFRA does not regulate the inert ingredients in pesticide preparations, although these could also have adverse effects. The purpose of inert ingredients in pesticide preparations is to make the active ingredient soluble, to stabilize the active ingredient, or to allow it to be applied in a specific way.

Integrated pest management (IPM)

Farmers commonly use pesticides according to a predetermined calendar schedule rather than on the basis of need. IPM is different from conventional agriculture in important ways: *A major tenet of IPM is that pests should be managed, not eradicated.* In IPM, pesticides are not automatically used. This is a very important component of IPM – instead of spraying according to a predetermined schedule, a farmer practicing IPM assesses fields to identify what pests are present and to check for signs that a pest is beginning to reach a stage where it can cause significant damage. Only then is pesticide applied. IPM can reduce pesticide use 50% to 70%, depending on the crop in question.

The practice of IPM also entails choosing pesticides with low toxicity to humans and other nontarget organisms and applying them in the lowest quantities that are effective. For IPM to work, farmers must thoroughly understand the crops they are growing, the pests that infest them, and the way climatic conditions affect both crops and pests. They must also be willing to make the effort to follow through on their knowledge. IPM is a form of P[2] because if less pesticide is applied, less pesticide can contaminate the environment or food.

IPM has other components: Eliminate pest-breeding places by destroying crop residues after harvest. Otherwise, pests may survive in the residues to attack the crop the following season. Time the planting of a crop to minimize its vulnerability to pests and the need for pesticide. Use mechanical cultivation to control weeds; that is, destroy them with hand tools or farm machinery. This reduces herbicide use but requires more human or machine effort. Grow crops in ecologically appropriate regions, locales that require the least pesticide for a successful crop; cotton, for example, grows well in some regions with little or no pesticide input, whereas in other regions it requires large amounts. Unfortunately, growing a crop in appropriate regions may not be easy or even possible for established farmers. As one example, a farmer may already have a fruit orchard growing in a region where pesticide use is necessary. With a large investment in long-lived trees, the farmer is unlikely to switch to a crop better suited to the region.

Crop rotation is growing different crops from one year to the next or from one season to the next. It is a useful means to control pests that cannot long survive unless a specific crop is present. For example, corn may be planted in one year and soybeans the next. Rotation can have other advantages. If legumes are planted as part of the rotation, they replenish soil nitrogen and organic matter so, when the alternate crop is grown, less pesticide and less fertilizer are needed. However, many farmers grow large tracts of monoculture corn, cotton, wheat, or other crops, which they are unwilling to rotate. They may, for example, have a heavy investment in machinery specific to the crop grown, which may make it difficult to switch crops. Intercropping, growing more than one crop on the land at the same time, is another technique to lower dependence on pesticides. Again, some farmers growing monoculture crops may not find intercropping reasonable, especially because it involves more labor.

EPA is working to enroll 75% of the country's agricultural acreage in IPM programs by the year 2000. In 1993, to assist in reaching this goal, EPA, FDA, and the USDA started a cooperative endeavor with businesses. Its intent is to develop means to reduce pesticide use and to develop safer methods of handling and applying pesticides. Corn growers are sponsoring a contest seeking means to maximize profits from growing corn while using less pesticide. California pear growers are carrying out farmer education programs to encourage alternative pest control strategies and research into safer use of pesticides. Growers are evaluating biological controls such as pheromones and insect growth regulators and making financial contributions to IPM research. Seventeen utility companies that use pesticides along power line rights-of-way are promoting worker training to lower the risks of pesticide use to humans and the environment.

Biological control. When a pest's natural enemies are known, biological control (biocontrol) can be useful. Most often, a pest's natural predator, a parasitic insect or a microorganism that causes a disease in the pest, is introduced to help control a pest. Examples of insect predators are praying mantises and ladybugs, which can be reared in large numbers and released to infested areas, where they, at least temporarily, reduce insect populations. Sometimes an exotic enemy is imported from a distant location; in this case, careful study is needed beforehand to assure that the exotic will not itself become a problem. Biocontrol cannot always fully replace pesticides. For example, parasites can partially control the alfalfa weevil, but some pesticide is still needed.

Another form of biocontrol uses chemicals produced by other organisms to control the pest. An example is pheromones, chemical attractants produced by insects. The pheromone produced by a female pest may be used to attract the males of the species into a trap. Another example is growth regulators, chemicals produced by insects that control stages of the life cycle. Sprayed on a crop infested with insect larvae of specific species, a growth regulator interferes with their development into adults. This use of chemicals from other organisms is referred to as *biocontrol*, but human intervention is necessary. In the examples given here, the pheromone or growth regulator must be purified from the insect and its structure elucidated. Then, because such tiny amounts of these chemicals are produced by the insects themselves, they must be synthesized to provide useful quantities. In effect, they become synthetic pesticides, but the information used to make them is from living organisms. Another form of biological control involves less complicated human labor: Large numbers of eggs from an insect pest are hatched, raised to adulthood, and sterilized; when released, the sterilized pests mate with normal insects, but no offspring are produced. Sometimes biological control simply means making sure habitat is available for a pest's natural enemies. See Table 10.2 for a summary of alternatives to synthetic pesticides.

More than 100 years ago an insect called the cottony-cushion scale became a major threat to the California citrus industry. This pest was not an American insect, but had been introduced from Australia. Australian entomologists were subsequently able to identify two enemies of the scale and, in the late 1880s, took them to California. Within a few months the scale was brought under control, and it remains under control. Nearly a hundred years after this early example of biocontrol, cassava plant roots (a basic food crop for millions in Africa) were infested by the cassava mealybug, an insect that grows entirely on cassava. In heavily infested fields, farmers lost up to 80% of their crop. The mealybug pest was not from Africa; but some detective work revealed that it was from South America, the same continent from which cassava came. Subsequently, a wasp, *Epidinocarsis lopezi*, a natural enemy of the mealybug, was

TABLE 10.2 Alternatives to conventional pesticides

ALTERNATIVE	DEFINITION	EXAMPLES
Natural	An extract prepared from living organism	An extract of chrysanthemum, tobacco, or the neem tree
Botanical	An ingredient purified from a natural extract	Pyrethrum purified from chrysanthemum extract or nicotine from tobacco extract
Biological [a]	A living agent such as bacterium, insect, or fish that acts as a pesticide	After rabbits were introduced into Australia, their population exploded; they were long controlled by a microorganism that infected and killed them
Microbial [a]	A biological that is a microorganism (bacterium, virus, fungus, or protozoan)	The Bt toxin or baculovirus, either of which can act as a pesticide
Bioengineered organism [b]	A living organism with genes from another organism inserted into its DNA, which allow it to make a pesticide that it was previously unable to produce	A plant with the Bt gene inserted into its DNA, allowing it to produce a product that repels several insects, or a plant given genes allowing it to make a product, making it resistant to herbicides

[a] These are not extracts or chemicals, but living organisms.

[b] In a *biological*, a living organism is the pesticide. In a *bioengineered organism*, the living organism produces the pesticide after it has been given the necessary genetic material to do so.

identified in South America. After careful study, the wasp was introduced into African cassava fields. Once it was clear that the wasp effectively killed mealybugs, researchers started a program to breed the wasp and distribute it over the large region of Africa affected by the mealybug.

The introduction of an exotic insect – from another region or country – has often led to the introduction's itself becoming a pest. This is because the insect may have no enemies in its new setting. The same is true of exotic plants and animals. That is why the South American wasp was carefully studied before its introduction into Africa. However, an insect that preys on other insects has a fortunate characteristic. It usually preys only on the specific species with which it evolved. Because it only attacks the pest it was introduced to control, the usual concern associated with introducing an exotic species is greatly reduced.

The Californian and African examples given here do not represent isolated instances. In more than 160 countries, about 560 biological control agents have been introduced against nearly 300 target insect pests; substantial or complete pest control resulted. Professor L. E. Ehler of the University of California at Davis points out that this remedy does not have the toxic side effects of chemical pesticides and is often permanent and economical. Usually, the biocontrol agents are readily available. Not only insects, but weeds and other pests, can be biologically controlled. As is the case with insects, many weeds that infest United States crops are exotics. To control an exotic weed, researchers go back to its original home to find an insect or pathogenic microorganism that can control it. As with insects, a weed enemy must be carefully studied to minimize the chance that it will itself pose new problems.

Organic farming

Organic farmers differ from farmers using IPM. They either use no pesticides at all or "natural" pesticides. However, they do use many of the techniques described earlier, such as intercropping and crop rotation. Organic farmers urge a return to farming without chemical pesticides or synthetic fertilizers. One problem with current organic farming is that it does not produce the huge amounts of foodstuffs we are accustomed to getting from high-input farming. Thus, more land is needed to grow the same amount of food. Critics of organic farming point out that as human population increases and progressively less farmland is available, we must find means to glean large crops from the same amount of land. Supporters of organic farming reply that although there are crop losses over a period of years as a farmer makes the transition to organic farming, the switch can be made and production can be as high as that obtained with pesticides. They also indicate that more research on organic farming is needed and that its potential is probably as great as that of the high-input agriculture now used. Furthermore, it would be sustainable over the long term. The USDA currently provides few research dollars for the study of sustainable agriculture. Between 1990 and 1995, it spent only $28.5 million on sustainable agriculture research, but $640 million on biotechnology.

Whatever agricultural system is used, some fear that as human population continues to increase, no agricultural system will be able to produce an adequate food supply. One obstacle to organic farming is social. Consumers want unblemished fruits and vegetables. How many of us would willingly buy peaches with blemished skin, apples containing wormholes, or broccoli with worms? Consumers need to be urged to care less about cosmetic defects or small amounts of insect or other pest damage; farmers could then use pesticides only for the urgent purpose of saving a crop.

DISCUSSION QUESTIONS

1. There is a major concern that modern agriculture is not sustainable.
   (a) What does this statement mean to you?
   (b) What are the reasons that modern agriculture may not be sustainable?
   (c) Compare organic farming and farming that uses IPM. Do you believe IPM can be sustainable? Explain.

THE SEARCH FOR ALTERNATIVES

Chemical companies are researching the development of pesticides that are
less toxic to humans and that work at much lower concentrations. Some new
herbicides are effective when applied at 0.02 lb/acre. Compare this to the
more typical 2 lb/acre for older pesticides. One company developed an herbicide that is applied at teaspoons per acre and, because it only affects a plant
enzyme, has very low toxicity to humans and animals. Another developed an
insecticide also only applied in tiny quantities that, because it only affects insect enzymes, shows low toxicity to nontarget species. One product recently
approved by EPA for use on corn and soybeans has an LD_{50} in rats of greater
than 500 mg /kg body weight. As can be seen in Table 3.4, this means that it is
only slightly toxic. In addition, it is not a teratogen or carcinogen, breaks
down quickly in the environment, and is not a threat to groundwater. As seen
in Table 10.3, these are all desirable traits. However, it is probably impossible
to incorporate all the characteristics noted in this table in one chemical. It is
also tremendously expensive to bring new pesticides to market, requiring 7 to
8 years to do so. These constraints lower the rate of desirable changes.

More generally, manufacturers are examining botanical pesticides, biological pesticides, and bioengineered products as potential new pesticides (see
Table 10.2). Pyrethrum, purified from chrysanthemums, is an example of a
botanical pesticide. Pyrethrum itself has limited applications, and more effec-

TABLE 10.3 Desirable characteristics in a pesticide

Only a small amount is needed to kill targeted pests

Has low toxicity to nontarget species

Is specific to one or a few pests

Has a lifetime only long enough to kill target pests (does not persist in environment)

Degrades into benign products

Does not bioaccumulate

Does not run off with water from application site

Pests are slow to develop resistance to it

Note: A pesticide would ideally have most of these characteristics. In practice, some
characteristics are incompatible. For example, it is desirable that a pesticide not migrate
from its application site. However, the reason it may not migrate is because it binds tightly
to the soil and is not soluble in water – characteristics often associated with the undesirable
traits of environmental persistence and bioaccumulation.

tive synthetic pyrethrins have been developed. Four insecticides now available in the United States are derived from the neem tree, grown in India. Like pyrethrum, neem pesticides have limited current applications because, when exposed to sunlight, they quickly degrade to products that cannot kill pests. However, it may be possible to overcome some of their limitations.

A microbial pesticide is a microorganism that has pesticidal properties and is, in fact, applied as a chemical pesticide. A well known microbial is the bacterium *Bacillus thuringiensis* (Bt) used to control caterpillars, beetles, and flies. Microbial insecticides such as Bt and baculoviruses are more pest-specific than chemical pesticides, lessening the threat to nontarget species. But, like other pesticides, a microbial must be shown not to threaten humans, animals, or the environment. Also, although pesticides specific to one pest are more desirable environmentally, they may not be profitable to the company that manufactures them because of this limited application.

Biotechnology companies are heavily investing in research aimed at developing bioengineered organisms. In a bioengineered plant, for example, specific genes are inserted into the genetic material to give it properties it did not previously have. Plants already modified by bioengineering include cotton and tomatoes that can resist an herbicide. When the crop is sprayed with that herbicide, only the weeds growing with the crop are killed, not the crop itself. This approach is controversial because some believe it encourages continued dependence on pesticides. Manufacturers respond that herbicide-resistant crops lessen the need for herbicide because it is applied when it can be most effective. For example, instead of applying herbicide before a crop emerges from the ground, previously necessary to prevent the crop itself from being killed, the herbicide can be more beneficially applied later. In a different application of bioengineering, genes are taken from the Bt bacteria mentioned above. After manipulation, these are transplanted into such plants as cotton, tomato, and potato. The plants receiving the Bt genes use the new genetic information to produce a protein that kills specific insects. Pests can develop resistance to Bt toxins, but there are many groups of Bt toxins. When resistance develops to one, it may be possible to replace it by a different one.

Some believe that bioengineering is unnatural. The response to this criticism is that plant breeders have always deliberately selected the traits and, in effect, the genes, they want to propagate. But whereas in traditional plant breeding the plants may need to be grown over many generations to obtain a desired trait, the insertion of genes allows the first generation to express it. But bioengineering can take more controversial approaches. For example, when the microbial pesticide baculoviruses is sprayed on a crop, 4 to 7 days elapses before the target insects are killed. This time can be speeded up by the following means: A scorpion gene that codes for a toxin is inserted into the baculovirus. When the baculovirus that can make the scorpion toxin is sprayed on plants, insects die twice as rapidly. This type of bioengineering, which takes genes from an exotic source like a scorpion and transplants them into a virus, leads to fears that a dangerous supervirus may be created. Even the use of a "simple" bioengineered product containing Bt genes, which come from a well-understood microbial pesticide, raises concerns that superresistant insects could develop.

Some third world countries have fewer qualms about going ahead with bioengineering projects, often without the careful evaluations considered crucial in the United States or other developed countries. There is increasing pressure to develop an international biosafety protocol to control releases of bioengineered organisms into the environment. Alternatives to synthetic pesticides are desirable for many reasons. But, as even this brief overview indicates, alternatives can also be very controversial. They cannot be, and are not assumed to be, problem-free. Developed countries, at least, will carefully scrutinize them before they are marketed. Pest species have had millions of years to develop survival strategies, and it is unlikely that humans will entirely outwit them.

DISCUSSION QUESTION

Is an international biosafety protocol needed to control the releases of bioengineered organisms into the environment? Why?

FURTHER READING

Agriculture and Pollution Prevention. 1995. *Pollution Prevention News*, EPA 742–95–001c, 3–6, May-June.

American Chemical Society. 1987. *Pesticides information pamphlet*. Washington, D. C.: ACS.

Gardner, G. 1996. IPM and the War on Pests. *World Watch*, 9(2), 20–27, April.

Gianessi, L. 1993. The Quixotic Quest for Chemical-Free Farming. *Issues in Science & Technology*, X(1), 29–36, Fall.

Hileman, B. 1995. Views Differ Sharply over Benefits, Risks of Agricultural Technology. *Chemical & Engineering News*, 73(34),8–17, 21 Aug.

O'Riordan, T., Clark, W. C., Kates, R. W., and McGowan, A. 1995. Health Report for Earth. *Environment*, 37(3), 11, April.

Pinholster, G. 1994. Debatable Edibles: Bioengineered Foods. *Environmental Health Perspectives*, 102(8), 636–39, Aug.

Smith, K. R. 1995. Time to "Green" United States Farm Policy. *Issues in Science & Technology*, XI(3), 71–8, Spring.

U.S. Environmental Protection Agency. 1995. *Citizen's Guide to Pest Control and Pesticide Safety*. EPA (7501C), EPA 730-K-95–001, Sept.

Woods, M. 1991. Nature Makes Its Own Toxins. *Chemecology*, 20(5), 12–13, July/ Aug.

11

ENVIRONMENTAL ESTROGENS

Estrogens are chemical hormones naturally produced by female animals. They are carried in the bloodstream to responsive tissues where they stimulate and maintain changes that make an animal female. They are also necessary in males, but in much smaller amounts. Agents that mimic natural estrogens are referred to as *environmental estrogens* or *xenoestrogens*. This chapter focuses on xenoestrogens; however the animal body produces many hormones in addition to estrogens. An environmental agent that can mimic one or more hormones is called an *endocrine disrupter*, a term much broader than *xenoestrogen*. The term *environmental agent*, is deliberately used because not all agents suspected of being xenoestrogens are chemicals.

POLLUTANTS AND THEIR PROPERTIES, SOURCES, AND REDUCTION

Chemicals or other environmental agents that show estrogenic properties can be either natural or synthetic. Many categories of chemicals and agents can mimic or partially mimic estrogens (see Table 11.1 and Figure 11.1). Some prominent xenoestrogens such as DDT are chlorinated chemicals; others, such as PAHs and mold toxins, contain no chlorine. Others such as lead, cadmium, and mercury are not even organic chemicals; rather they are metals. A number of pharmaceuticals and other drugs also have estrogenic properties. So does at least one virus, the human papilloma virus discussed in Chapter 4. Because additional estrogenic chemicals continue to be identified, the answer to the question, What are the pollutants of concern? is far from complete.

Pesticides

As of late 1993, 35 of the 45 chemicals identified as having estrogenic properties were pesticides. The insecticide DDT, widely used from 1943 until 1973, is

245

TABLE 11.1 Agents with estrogenic properties and examples of each

INDUSTRIAL CHEMICALS	DRUGS
Pesticides (DDT, kepone, dieldrin)	Stomach ulcers (cimetidine)
Dioxins and furans	High blood pressure
PCBs	Birth control (estrogens)
PAHs (benzo[*a*]pyrene)	"RECREATIONAL" DRUGS
Plastic degradation products (bisphenol)	Alcohol
Metals (cadmium, lead, mercury)	Tetrahydrocannabinol (in marijuana)
NATURAL CHEMICALS	OTHER AGENTS
Phytoestrogens	Human papilloma virus
Mold toxins	EMFS (suspicion)

the best known of these. DDT was also the first chemical implicated as having adverse effects on animal reproduction; in some bird species, including bald eagles and osprey, it thinned egg shells; it also affected reproduction in bird species whose egg shells were not affected. A major population drop seen in Western sea gulls was attributed to DDT exposure. Many so-called lesbian

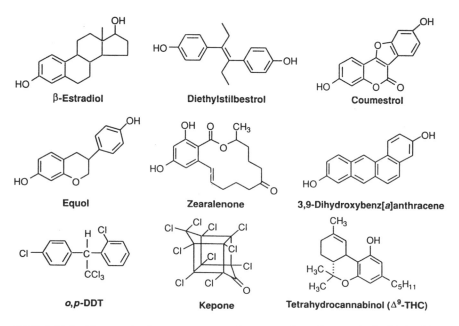

FIGURE 11.1. Chemicals showing estrogenic properties. *Source*: J. A. McLachlan, NIEHS, Research Triangle Park, North Carolina, 1993.

gulls shared nests with other females and produced eggs that contained males with feminized reproductive tracts; male gulls showed feminine characteristics and some were sterile. DDT was banned in 1973, and although some bird populations may still be affected, many have recovered. Another major example of adverse effects on an animal population occurred after a spill of the pesticide dicofol into Florida's Lake Apopka in 1980, which later resulted in the lake becoming a Superfund site. The dicofol contained about 15% DDT as a contaminant. Later, it was observed that 95% of Lake Apopka's alligator eggs were failing to hatch, as compared to a normal hatch rate of 70–80%. Half of those that did hatch died within 2 weeks and those that survived had feminine characteristics. These effects were traced to *p,p'*-DDE, a degradation product of DDT. DDE is not a "true" xenoestrogen because, rather than acting as an estrogen, it blocks the action of androgen, the male hormone.

Pesticides with estrogenic properties have low potency as compared to natural animal estrogens. The effects on birds and alligators noted earlier occurred in the presence of high doses. The high DDT concentration associated with eggshell thinning resulted from its buildup in the environment over the 30 years in which it was used, and the high DDE concentrations in Lake Apopka resulted from a spill. Since DDT was banned in 1973, its levels in the American environment and in animal and human blood and tissues have declined, sometimes dramatically. The environmental concentrations of other polychlorinated pesticides that, like DDT, have been banned are also low. Nonetheless, a 1996 report was disturbing because it raised the possibility of synergistic effects. Recall from Chapter 3 that the net effect of more than one chemical can be antagonistic (the chemicals cancel or partially cancel out each other's effects), additive (the effects add together), or synergistic (the effect is greater than additive, sometimes much greater). The 1996 report presented data showing that any one of four chlorinated pesticides that were tested had only weak estrogenic properties. However, when two were tested together, the effect was synergistic. This finding enhanced concerns that xenoestrogen could have effects even at low environmental levels. In addition, one of the four pesticides tested, endosulfan, is still used in the United States.

As seen in Table 11.1, various chemical families show estrogenic properties. A number of these will be discussed. Incorporated into the discussion of each are the answers to the questions, What are the pollutants of concern? Why are they of concern?, and so on.

Polychlorinated biphenyls (PCBs)

Another group of chemicals implicated in reproductive problems is PCBs, a family of polychlorinated chemicals (see Box 3.3). By the time they were banned in the 1970s, 2 to 3 billion pounds of PCBs had been produced and large amounts released onto land or into water. Many sites around the Great Lakes were among those highly contaminated with PCBs. Fish in or near these sites were also highly contaminated. In one study, Dr. John Giesy of Michigan State University fed Great Lakes fish (that contained high PCB concentrations) to laboratory-raised mink. When the fish constituted 20% of their diet, the mink had a reduced number of kits with a reduced survival rate; when the fish were 40% of their diet, the mink did not reproduce at all. In the twenty years since PCBs were banned, their concentrations in Great Lakes fish have dropped as much as 90%, although sensitive animal species such as mink still suffer. Another example of adverse effects attributed to PCBs occurred in terns nesting off the Massachusetts coast. These birds attracted attention because of the abnormal behavior of nesting females. As had occurred with gull populations exposed to DDT, female terns were found sharing one nest. Investigation revealed that some male embryos in the eggs of these terns had both male and female tissue, or improperly developed male organs. Their nesting area was near a site highly contaminated with PCBs.

The greatest route of human exposure to PCBs is fish consumption. Although PCB concentrations in Lake Michigan fish have dropped greatly over the years, they remain high enough that advisories on their consumption are issued. Concern arose as to whether PCB levels in these fish could affect human reproduction. In a study designed to examine this concern, children born to women who ate 2 to 3 Lake Michigan fish a month were studied. The babies of these women were reported to have low birth weight and, at age 4, to exhibit short-term memory problems. The design of this study was criticized and its reported results remain controversial. However, human populations have clearly shown adverse effects in response to higher doses. In the late 1970s, in Japan and Taiwan, thousands of individuals were poisoned after ingesting rice oil accidentally contaminated with PCBs. Women who were pregnant at the time had an abnormal number of fetal deaths and stillbirths. Babies born to women affected by the Taiwanese incident had low birth weight, abnormal skin pigmentation, and erupted teeth. They later showed lower IQ scores than a control group of children and more behavior problems.

Although environmental levels in the United States have greatly decreased, PCB levels in human blood and tissues have declined more slowly than those of DDT. In fact, environmental levels of PCBs appear to have reached a plateau and are now declining much more slowly than previously. Although the process is expensive, PCBs in land-based hazardous waste sites can be cleaned up. It is more difficult and expensive to clean up the contaminated sediments where PCBs, because of their low water solubility, concentrated. In sediments, they are sheltered from the sunlight and oxygen that could assist in degrading them; even in sediments, microorganisms slowly break them down.

Dioxins

Dioxins are a family of chlorinated chemicals related to PCBs. They have powerful biological effects and can interfere with the function of a number of hormones. Animals are particularly sensitive to dioxin exposure when it occurs in the uterus or shortly after birth. In one-well known study, pregnant rats that were given one tiny dose of the most toxic dioxin, 2,3,7,8-TCDD, produced male offspring that demonstrated reduced sperm production and other adverse effects. However, unlike PCBs and DDT, dioxins have antiestrogenic properties in some situations, and some less toxic dioxins are even being evaluated as potential treatments for estrogen-sensitive breast cancers. Considering their extreme toxicity, it is fortunate that environmental levels of dioxins are far lower than those of other polychlorinated chemicals. This is because, excepting tiny amounts synthesized in research laboratories, dioxins are not deliberately produced.

Dioxins form as contaminants during combustion, especially combustion of MSW and medical waste. Forest and grass fires also generate dioxins, as do other combustion sources. Once emitted to the atmosphere, the dioxin particulates drift onto water and land. They settle onto vegetation that is eaten by cattle and other animals and concentrate in the animals' fat and milk. Dioxins also form as contaminants during the chlorine bleaching of wood pulp used to make paper. Released in bleaching effluents into rivers, dioxins (like PCBs) concentrate in sediments, from which they are taken up by invertebrate animals that are subsequently eaten by fish. Birds and mammals, including humans, are exposed when they eat contaminated fish. Except for people who eat larger than average amounts, fish account for about 7% of total dioxin exposure. Most of the remaining exposure results from consumption of meat and dairy products. A vegetarian diet can lower dioxin intake by about 98%. In the homes of smokers, burning cigarettes emit dioxins, tiny in amount, but enough to double the smoker's dioxin exposure. Dioxins are also generated by home wood burning and accumulate in wood soot.

MSW and medical waste combustors are the major known sources of dioxin emissions, although their emissions vary greatly: Dioxin amounts released by medical waste combustors varies from 117 to 450 nanograms per cubic meter (ng/m^3), and that released by MSW combustors from 1 to 10,700 ng/m^3. New EPA regulations should reduce the emissions of particularly strong dioxin emitters up to 99%. It is useful to remember that although combustion is the source of 95% of dioxins in the environment, properly controlled combustion can be and is used to destroy dioxin-contaminated materials. Dioxins were also previously formed as contaminants during the manufacture of chlorinated pesticides, but modified processes have greatly reduced their formation; in some cases, the pesticides were banned. The process used to bleach wood pulp with chlorine has also been modified since the mid-1980s, resulting in reduction of dioxin discharges to water of up to 95%. Even at the tremendously sensitive detection level of 10 parts per quadrillion, 2,3,7,8-TCDD is not ordinarily detected in bleaching effluents. In mills that switch from elemental chlorine to chlorine dioxide as a bleaching chemical, effluents do not contain detectable dioxin. Some call for elimination of any chlorine-containing chemical in pulp bleaching with the goal of removing any remaining dioxin emissions, detectable or not. Presumably as a result of increased controls placed on dioxin-emitting sources, dioxin contamination of the U.S. environment has decreased by about half in the past 20 years.

Polycyclic aromatic hydrocarbons (PAHs)

The PAH family of chemicals is related to PCBs and dioxins, but does not contain chlorine. Some are known human carcinogens and some show estrogenic properties. PAHs are also environmentally persistent substances that bioaccumulate in animal fat. Like dioxins, PAHs are produced by almost any combustion process. However, whereas dioxins are primarily produced during waste combustion, the major source of PAHs is fossil fuel combustion although some are formed in wood-burning fireplaces and stoves. Within the home, cigarette smoking can be a major source. Charcoal-broiled foods especially, but also oven-browned foods and toast, contain these chemicals. PAHs cannot be eliminated from the environment, but they can be controlled. PAH emissions from coal burning are down as compared to historical releases, and the same is true of motor vehicle emissions. PAHs produced in the home can be reduced by preventing grease from dripping onto charcoal-broiled foods and preventing fried or baked foods from becoming too brown. Fireplaces and stoves need regular maintenance to minimize soot formation and emissions. Smoking cigarettes within the home can be eliminated. However, unlike PCBs, polychlorinated pesticides, and dioxins, which society either has eliminated or strictly controls, PAH emissions – although reduced – continue as a result of our ongoing dependence on fossil fuels.

Natural estrogens

Plant xenoestrogens (phytoestrogens) in the diet may protect against the development of breast cancer, osteoporosis, and menopausal symptoms. However, as is the case with industrially-generated xenoestrogens, phytoestrogens can also adversely affect sexual development and reproduction. In the 1930s, Australian sheep suffered a sharp fertility decline and displayed other symptoms as well; an investigation revealed that the responsibility rested with the particular clover being grazed by the sheep, one producing the phytoestrogens equol and coumestrol. Phytoestrogen interference with cattle fertility has also been observed. In laboratory experiments, rats fed the phytoestrogen coumestrol showed premature estrous cycles and delivered young that showed impaired sexual development. The finding, in a soy-based infant formula used in New Zealand, of 3 to 5 times as much of the phytoestrogens daidzein and genistein as needed to disrupt a woman's menstrual cycle posed a more direct concern. Whether the infants were adversely affected was unknown, but the study indicated the need for an investigation. Even corn oil and other polyunsaturated fats show estrogenic effects. The chemical tetrahydrocannabinol, found in marijuana, has estrogenic properties. So do mycotoxins produced by fungi growing on food crops. The human diet contains very high amounts of phytoestrogens as compared to industrial xenoestrogens, but some believe they will prove to be benign or beneficial. They point out that Asians have eaten a high-soybean diet for hundreds of years. Others note that Asians may have eaten soy for so long that any sensitive individuals whose reproduction was impaired have been bred out of the population. It is believed that plants have developed chemicals that affect animal reproduction as a means to keep predators that prey on them in check. There is a general acceptance that phytoestrogens must be researched along with industrial xenoestrogens.

Other estrogenic chemicals

- A number of chemicals associated with plastics and plastic use have estrogenic properties and are pervasive in the environment. Cans coated inside with a plastic resin (to prevent a metallic taste) leak the weakly estrogenic chemical bisphenol-A into the food. Bisphenol-A also leaches from a resin used as a sealant to protect children's teeth against decay. Phthalates, chemicals used to make plastics flexible, leach from plastics. Nonylphenols are also shed by plastics. The chemical DBP leaches from certain plastic wraps used on foods and is also found in plastic plumbing pipes.
- The weakly estrogenic preservative BHA is an antioxidant used as a food additive.
- The metals mercury, cadmium, and lead show estrogenic activity.
- Some drugs prescribed to lower high blood pressure have estrogenic properties. One antiulcer drug, cimetidine, has been associated with breast growth in some men.

- Sewage effluents contain chemicals with estrogenic activity. In a 1995 article, researchers reported testing 20 chemicals in sewage effluent, chosen at random from among the thousands of chemicals present. Half were weakly estrogenic.

BOX 11.1

Exotic xenoestrogens? Two aspects of our use of electric power may affect lifetime exposure to estrogens. One of these is bright light at night. Modern girls reach puberty years earlier than was the case historically, a fact that puzzles scientists. Some believe that earlier puberty is due to better nutrition. Others suspect that xenoestrogens may play a role. Yet others wonder whether an artificially lengthened day – bright light at night – is a factor. The day lengthening artificially produced by bright light reduces the release of the hormone melatonin by the pineal gland (a small gland in the brain that can detect the difference between day and night). The pineal only releases melatonin in the dark. This fact is significant because melatonin is an *anti*estrogenic chemical. Regular exposure of immature girls to bright light may lessen melatonin release, which could enhance estrogen production in girls and stimulate earlier puberty. Over many years, bright light at night may also increase a woman's lifetime exposure to estrogen by inhibiting the release of the aniestrogenic melatonin, and thus increase breast cancer risk. This is a fascinating hypothesis, but testing it has barely begun.

Another aspect of electric power use has garnered a great deal of publicity: This is the possibility that exposure to EMFs (the electric and magnetic fields associated with anything that carries electricity or operates with electricity) may have adverse effects. We are all exposed to EMFs from power lines and, especially, from the many appliances and other electronic devices in our homes and work places. Some epidemiologic studies have indicated that high exposure to EMFs is associated with an increased risk of breast and other cancers. EMFs also stimulate the growth of breast cancer cells that are cultured in the laboratory. However, an NRC panel – after a 3-year research project that analyzed the many EMF epidemiologic studies – reported in late 1996 that it found no evidence of an adverse effect of EMFs at levels found in human residences. Despite this finding, concerns about EMFs are unlikely just to go away. As epidemiologist Dimitrios Trichopoulos of Harvard University was quoted as saying in a November 1996 *Science* article, "It's one thing to say 'not guilty', and another to say, 'innocent'."

XENOESTROGENS AND HUMAN HEALTH

It is probable that hundreds, likely thousands, more chemicals with estrogenic or other hormonal properties will be identified. Those already identified are found everywhere in the environment, including food, water, and air. The significance of these chemicals is not clear, and the major and controversial questions now being asked are: Can xenoestrogens, at levels found in the environment, adversely affect humans? Do they cross the placenta and reach the fetus in amounts that could adversely affect reproductive development? Do they contribute to the development of breast cancer in women or testicular cancer in men?

Women

Estrogens are necessary for normal development in female animals. At the same time, a woman exposed to larger than average doses of even her own body's estrogen over a lifetime has an increased risk of breast and certain other cancers. Some natural biological factors contribute to the risk. Women with early puberty or late menopause receive a greater lifetime dose of estrogens and have a greater breast cancer risk. Life-style also affects breast cancer risk. Even hundreds of years ago, it was noted that nuns had breast cancer more often than married women. In twentieth-century America, it is known that childless women, women who bear their first child at a later age, and women who do not breast-feed their infants are at greater risk. It is known that alcoholic women have enhanced breast cancer rates. Other factors that may increase risk are cigarette smoking and, possibly, eating of red meat. Because its overall beneficial effects on health are so great, estrogen replacement therapy is often used by postmenopausal women although that therapy exposes them to a higher lifetime dose of estrogen. Exercise appears to lower risk; one well-regarded study found that women under 40 who exercise at least 4 hours per week lowered their breast cancer risk 60%, and even 1 to 3 hours per week reduced risk 30%.

Some scientists believe that the risk factors noted here only partially explain the increased breast cancer rate seen in industrialized countries. Whereas an American woman's lifetime risk of breast cancer in 1960 was 1 in 20, today's risk is 1 in 8. Risk increases with age: Relatively few breast cancers occur before age 50, but by age 65 the risk is 1 in 16 and by 85, 1 in 8. (One, perhaps dubious, comfort is that, although the U.S. breast cancer rate is up, the death rate from breast cancer is down.) Some suspect that some of the higher risk is due to xenoestrogen exposure. One particular suspicion is that older women built up high levels of DDT in their bodies in the years before it was banned and, because of this, may have a higher breast cancer risk; a 1993 New York study seemed to lend support to this idea. It obtained results showing that women with breast cancer had higher blood levels of DDT and of its breakdown product, DDE, than did those without breast cancer. Results of a 1994 study did not support such an association; this second study examined frozen blood samples that had been taken from women having physical examinations in the 1960s, when DDT was still in use. Three hundred blood samples, half from women who later had breast cancer and half from women who did not, were analyzed for DDE and PCBs. No association was found between DDE levels and breast cancer, although some results were puzzling. The same study showed no link between PCBs and breast cancer. In a presentation on breast cancer risk factors presented in December 1994 before a National Institute of Environmental Health Sciences audience, the epidemiologist Dr. Barbara Hulka referred to exposure to DDT, PCBs, EMFs, by-products of char-broiling food, phytoestrogens in soy foods, and breast implantation as "controversial suspected risk factors." She emphasized measures women can take to lower breast cancer risk, including controlling alcohol consumption, avoiding smoking, exercising regularly, breast-feeding infants for 6 months to a year or more, and, for postmenopausal women, losing weight.

In addition to breast cancer, endometriosis is another disease whose incidence is rising in women. Again, xenoestrogens are suspected as a possible cause. In endometriosis, cells that normally line the inside of the uterus are found in other locations in the body. This causes painful symptoms and can lead to infertility.

Men

The potential effects of estrogen-mimicking agents in male animals may be of greater concern because males normally produce only small amounts of estrogens, and thus environmental exposures may represent disproportionately higher doses than in women. Not only chemicals with estrogenic properties, but those with androgenic and antiandrogenic activity must be considered. A number of problems either are increasing in human males or may be increasing. One major concern is that environmental estrogens are associated with declines in sperm production. A 1992 European study concluded that sperm production had fallen by 50% in the previous 50 years and later studies appeared to confirm this decline. However, an investigation of sperm contributed to sperm banks by American men between the years 1970 and 1994 showed no decline in either sperm number or ability to swim. Yet another study released in 1996 reanalyzed an earlier study that had reported a worldwide fall in sperm production but did not find any sign of decline. Other investigators find fault with all the studies on sperm production done to date and say that we do not know whether sperm production is increasing, decreasing, or staying the same.

Another concern is prostate cancer. The rate of this cancer has apparently doubled in the past 10 years in some countries. However, the increase may be more apparent than real. The Harvard epidemiologist Professor John Bailar believes reported increases in prostate cancer are "largely or entirely spurious." He points out that improved medical techniques have allowed detection of tumors that would have gone unnoticed before because most prostate tumors grow very slowly. Indeed, up to 90% of men may have prostate cancer if they live long enough. Testicular cancer, especially in young men, is one cancer definitely known to be increasing: There has been a 2- to 4-fold increase in testicular cancer in the United States, Britain and Denmark over the past 30 years. Although rates are still low, the trend is disturbing. Other adverse ef-

---

BOX 11.2

Diethylstilbestrol (DES). With the exception of actual poisoning cases, the adverse effects of DDT and PCBs that have been observed occurred in wild animals or birds, or in laboratory animals. However, the adverse effects of the pharmaceutical DES were seen in humans. DES, an estrogen prescribed to pregnant women between 1948 and 1971 to prevent miscarriages, is 5 to 10 times more potent than the body's own estrogens. About 1 in every 100 of the daughters born to women taking this drug had vaginal cancer as a young adult. The daughters also had an increased risk of breast cancer and many had reproductive tract abnormalities that made them sterile. About 15% of the sons born to these women had genital malformations at birth, although most were minor. As men, the sons had lower sperm counts, more abnormal sperm, and an increased risk of testicular cancer. Despite these problems, a 1995 study found that the fertility of DES-exposed sons, even of those with genital defects, was not adversely affected. Sperm counts were lowered, but fertility was not. Results of this study support the viewpoint of those who are less concerned about environmental estrogens. They argue that if even exposure to this strong synthetic estrogen did not affect male fertility, then we need not be overly concerned about chemical pollutants whose potency may be much lower than that of natural estrogens. However, these sons had other adverse effects and many questions remain.

fects known to be increasing are undescended testicles and urinary tract abnormalities.

SCIENTIFIC UNCERTAINTY AND FUTURE DIRECTIONS

The scientific study of any subject is an ongoing story and there is always at least some uncertainty. As new information emerges, viewpoints must often be modified. A controversial topic raises special problems because added to the uncertainty are the sometimes strongly held beliefs that some individuals hold even early in an investigation. However, there is one statement that it is safe to make: Possible links between industrial xenoestrogens and human disease are taken very seriously by the scientific community, even by skeptics. Still, skeptics emphasize points such as the following: Estrogens in birth control pills do not obviously increase breast cancer risk and, in fact, decrease ovarian cancer risk. This is true although they are ingested in larger amounts and have greater estrogenic potency than do chemicals such as the pesticides DDT and endosulfan. Furthermore, we ingest large amounts of naturally occurring plant xenoestrogens in our food.

Conversely, scientists like Dr. Devra Davis of the World Resources Institute believe that xenoestrogens may contribute to increased breast cancer rates by promoting the formation of "bad" estrogen. There are two pathways that the body can use to break down natural estrogen. The product of one path is "good" estrogen, which does not seem to promote breast cancer, whereas the product of the other is "bad" estrogen, which may. Some studies indicate that xenoestrogens may stimulate the formation of bad estrogens. Concerned scientists also note that xenoestrogen effects may be cumulative; that is, if all xenoestrogen exposures are added together, our exposure may be significant. There is also a disturbing possibility that these chemicals may act synergistically as noted earlier.

On the other hand, skeptics such as Dr. Stephen Safe of Texas A&M University reply that people are exposed not only to xenoestrogens, but to chemicals that are antagonistic to xenoestrogens. The antagonists may cancel out the effects of the xenoestrogens. Furthermore, we are not only exposed to anthropogenic sources of xenoestrogens, but to naturally occurring phytoestrogens as well. Safe estimates that dietary exposure to phytoestrogens is about 400,000 times greater than exposure to industrial xenoestrogens: Whereas a 16-year-old American male consumes about 2.5 µg/day of estrogenic pesticides, he consumes up to 1,000 mg/day of phytoestrogens. But those who argue with the skeptics point out that phytoestrogens are rapidly degraded, whereas the polychlorinated DDT and PCBs are not. However, most polychlorinated chemicals have been banned in the United States, and more recently identified organic xenoestrogens have structures that should allow them to break down more readily in the body and the environment.

A hormone such as 17β-estradiol, a human estrogen, is active at tiny concentrations. This is true because it reacts with a specific receptor. Once stimulated, the receptor magnifies the hormone's effect by initiating a chain of reactions that can have far-reaching consequences. This ability of a hormone to act at extremely low doses makes the finding of environmental estrogens disturbing. However, as with most other chemicals, a hormone has a threshold (a dose below which adverse effects are not seen, as described in Chapter 3). During the years that physicians prescribed DES to pregnant women, the Mayo Clinic prescribed low doses and no adverse reproductive effects were seen in the children of these women. However, another clinic prescribed doses more than ten times higher and adverse reproductive effects did occur in the children of women taking these doses. However, at yet higher concentrations in a living animal, a negative feedback comes into play and turns off the system. From this perspective, the truism "The dose makes the poison" is true only up to a point for hormones.

One major problem in evaluating the possible effects of endocrine disrupters, including xenoestrogens, is the difficulty of meaningfully evaluating chemicals for hormone-mimicking behavior. Consider a statement made by the University of Florida's Dr. Louis Guillette quoted in *Science News*, July 15, 1995: "We now know that no one technique, assessment, or species will be able to tell us whether an ecosystem is polluted" (that is, polluted with endocrine disrupters). Recall from Chapter 4 that, although imperfect, protocols exist to test a chemical for its ability to cause cancer and several other types of toxicity. For reasons beyond the scope of this text, evaluating potential endocrine disrupters is even more difficult. Nonetheless, future evaluations of a chemical's toxicity will need to include an assessment of its ability to act as an endocrine disrupter. An EPA panel is now working to develop a test strategy for those that pose the greatest threat.

To provide an overview of the whole endocrine disrupter field, an NRC panel started a study of endocrine disrupters and their potential effects on the hormonal balance of humans and animals in 1994. The panel is assessing the scientific literature on chemicals that have hormonal properties to characterize their known adverse effects and the problems that could result from exposure to them. It will identify human or animal populations that may be especially vulnerable. The fetus, for example, is especially susceptible to chemicals that interfere with sexual identity or development. The final NRC report will make recommendations on further research and monitoring needs.

A White House Task Force is also making an inventory of all federally funded research in the field. One spokesperson was quoted as saying, "There is a serious research need to find out if the alleged decrease in sperm counts and quality as well as if the increases in testicular, breast and prostate cancers are real or if they are reflective of our better ability to detect those diseases" (*Environmental Science & Technology*, June 1996, p. 242). EPA is meanwhile developing a research strategy on endocrine disrupters. Several other federal agencies and the chemical industry are also funding an increasing number of research projects on known or suspected endocrine disrupters. Betty Hileman, writing on the hypothesis that endocrine disrupters may be affecting people, commented, "In the popular press, the hypothesis is often characterized as either a confirmed scary reality or chimeric science. In fact, the truth probably lies somewhere between these two extremes, say most experts in the area." One point is clear. The information that has emerged on xenoestrogens and other chemicals with hormonal activity is complex and may become more so before answers are sorted out of the confusion.

DISCUSSION QUESTIONS

1. Are you satisfied with the actions being taken to deal with xeno-estrogens and other endocrine disrupters, or do you believe that more needs to be done? Explain.

2. Review the factors (Chapter 4) that make a pollutant a high priority. Considering these factors, what environmental estrogens (or categories of environmental estrogens) do you consider of most concern and why?

3. (a) Are you concerned enough about environmental estrogens to cut your consumption of alcohol?
   (b) What about marijuana, if you use it?
   (c) What about char-broiled foods?
   (d) If you are a woman, to what extent are you willing to change your life-style to lower your breast cancer risk?

4. Does the hypothesis that bright light at night affects reproductive development and, possibly, breast cancer development strike you as plausible? Explain.

5. The sewage effluent released by a community's wastewater treatment plant contains chemicals with estrogenic activity. It may contain hundreds.
   (a) If you were in the position of needing to evaluate this finding, and explaining it to your community, what would you say?
   (b) What additional information would you like to have?

FURTHER READING

Davis, D. L., and Bradlow, H. L. 1995. Can Environmental Estrogens Cause Breast Cancer? *Scientific American*, 273(4), 166–72, Oct.

Hileman, B. 1996. Environmental Hormone Disrupters Focus of Major Research Initiative. *Chemical and Engineering News*, 74(20), 28–31, 34–5, May 13.

Johnson, J. 1996. Endocrine Disruptor Research Planned by White House, Agencies, Industry. *Environmental Science & Technology*, 30(6), 242A-3, June.

Larkin, M. 1995. Estrogen, Friend or Foe? *FDA Consumer*, 29(3), 25–9, April.

Melius, J. M. 1995. Xenoestrogens and Breast Cancer Risk. *Health & Environment Digest*, 8(10), 77–9, Feb.

Raloff, J. 1994. Gender Benders: Are Environmental Hormones Emasculating Wildlife? *Science News*, 145(2), 24–7, Jan. 8.

Raloff, J. 1994. That Feminine Touch. *Science News*, 145(4), 56–9, Jan. 22.

Safe, S. H. 1995. Assessing the Role of Environmental Estrogens in Human Reproductive Health. *Health & Environment Digest*, 8(10), 79–81, Feb.

Wakefield-Alberts, J. 1995. Making Headlines in the Lab. *Environmental Health Perspectives*, 103(6), 560–2, June.

12

ENERGY PRODUCTION AND USE

The production and use of energy are major sources of pollution in the United States and worldwide (see Table 12.1). Americans use about one-quarter of the world's energy; Figure 12.1 provides information on the energy use of the average American as compared to that of another industrialized country, Japan, and that of several very poor countries. About 85% of the energy used in the United States is furnished by fossil fuels. By the year 2025, it is projected that worldwide demand for fuel will increase by 30% and demand for electricity by 265%. Thus, we must be concerned not only with the pollution resulting from current energy consumption, but from continually increasing consumption. We must find means both to conserve energy and to convert to less polluting forms of it. Otherwise, as one energy scientist, Dr. John Holdren of the University of California, has stated, "We're going to run out of environment before we run out of energy." Fossil fuels are both the most polluting and most heavily used form of energy. This chapter focuses on two major users of fossil fuel, motor vehicles and electric power plants. P² solutions are discussed as a means to reduce pollution resulting from energy production and use. The limitations of P² as a concept are also noted. For example, how do we compare fossil fuels that generate much pollution to energy sources that generate little pollution but produce other forms of environmental damage?

DISCUSSION QUESTIONS

1. (a) Does the statement, "We're going to run out of environment before we run out of energy?" strike you as realistic or unduly pessimistic?
   (b) If you believe it is realistic, what should society be doing differently?
   (c) What should you and I be doing differently?
2. The United States uses about a quarter of the world's energy. Each American uses about the same amount as 3 Japanese, 38 Indians, or

TABLE 12.1 Air pollutants and fossil fuel combustion

CRITERIA AIR POLLUTANTS	SOURCE
Carbon monoxide	In urban areas, up to 90% is from gasoline combustion in motor vehicles.
Nitrogen oxides [a]	High temperature combustion is the major source, especially in motor vehicles. Power plants are a secondary source.
Sulfur dioxide [a]	Electric power plants burning coal is the major source. Also formed during burning of any sulfur-containing fuel.
Ozone	Motor vehicles are a major source of the nitrogen oxides and VOCs, which form ozone.
Particulates	Fossil fuel combustion is a major source of the particulates of most concern.
Lead	Coal and gasoline combustion are common sources.
METALS	Coal and gasoline combustion are common sources.
VOCs	Gasoline combustion is a major source of hydrocarbon VOCs
GREENHOUSE GASES	
Carbon dioxide	Coal burning is predominant source. Petroleum combustion is also a source.
Methane	Agriculture is the major anthropogenic source, but methane is released during recovery of coal and petroleum deposits. It also escapes from pipeline leaks.
PAHs	These are formed in any combustion process (especially combustion in motor vehicles).

[a] Sulfur dioxide and nitrogen oxides are also the major acid rain precursors.

531 Ethiopians. In addition to environmental impacts, do you believe there are other concerns we should have about the amounts of energy the United States uses?

MOTOR VEHICLES

The United States has about a third of the world's motor vehicles, 140 million cars and 45 million other motor vehicles. These consume half of the petroleum used in this country; by themselves, motor vehicles use 8% more oil than the United States produces – If we did not burn gasoline in motor vehicles, we would not need to import petroleum. In addition to gasoline's heavy environmental price, our heavy dependence on foreign oil is considered a threat to

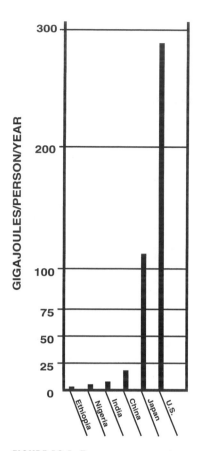

FIGURE 12.1. Energy consumption.

national security. Nonetheless, gasoline use continues to increase: Just between 1994 and 1995, gasoline consumption by Americans went up 4% as a result of an increasing number of larger vehicles, small trucks and minivans, that require more gasoline per mile driven. Motor vehicles present a pollution problem that sometimes seems insoluble. That situation may be changing.

What are the pollutants of concern, and why are they of concern?

Motor vehicle pollutants include carbon monoxide, nitrogen oxides, and VOCs. VOCs from motor vehicles are primarily hydrocarbons (chemicals containing only hydrogen and carbon). Motor vehicles also produce particulates that contain PAHs and metals (again, see Table 12.1). In 1990, motor vehicles emitted 65% of the carbon monoxide, 33% of the nitrogen oxides, and 29% of the VOCs. In urban areas, these percentages are higher. Using San Francisco as a representative city for the year 1990, vehicles emitted 75% of the carbon monoxide, 58% of the nitrogen oxides, and 38% of VOCs. Motor vehi-

---

BOX 12.1

A billion cars. A 1995 *World Watch* article chronicled the many ways that motor vehicles affect our environment and our lives. In 1950, there were 2.6 billion people in the world and 50 million cars; by 1995, those figures were 5.7 billion people and 500 million cars. By the year 2020, an estimated 1 billion vehicles will be on the road worldwide. Even in 1995, motor vehicles accounted for a third of global, and 50% of American, oil consumption. Tailpipe exhaust is the single largest source of air pollution in cities. There are the medical costs of diseases caused or aggravated by exhaust fumes. Also count in the more than 2 million motor vehicle accidents a year in the United States with more than 40,000 fatalities and the many serious physical injuries of those who survive accidents. Worldwide, nearly 500,000 people a year die in motor accidents.

There is also the loss of arable land to road building, already calculated to amount to 2% of total land in the United States or 10% of arable land. Private motor vehicles are also a major contributor to the urban sprawl that causes much more land to be lost to development. The World Resources Institute has calculated that about $300 billion a year of public funds is spent in the United States for road construction and repairs, routine street maintenance, traffic management, and parking enforcement. Traffic congestion leads to a good deal of time lost from the lives of many commuters. There are, of course, also the cost to individuals of purchasing, operating, and maintaining a car. One question that has been raised is, At what point will quality of life be so adversely impacted that Americans will opt for a life-style less dependent on cars and trucks?

---

cles also generate about 20% of the carbon dioxide emitted in this country as each vehicle produces its own weight in carbon each year. The concerns associated with these pollutants have been discussed in earlier chapters.

How is motor vehicle pollution being reduced?

Many of the steps now being taken to reduce motor vehicle pollution were mandated by the 1990 CAA amendments, the Clean Car Initiative, and the Energy Policy Act of 1992. With 185 million vehicles, the problem is immense and there is no one solution. Rather, many approaches are being used.

Conservation. As always, conservation is the ideal. Encouraging people to car pool, providing incentives to employers to get employees to car pool, and providing bike paths to encourage biking are some of these. Perhaps most important in the long run is building cities in a way that makes walking, bicycling, and use of public transport convenient.

Inspection and maintenance. About 10% of motor vehicles are responsible for 50% of all emissions. This 10% includes not just old vehicles, but vehicles of all model years that have not been properly maintained. In some locales that are out of compliance with the federal ozone standard, there are testing programs to identify offending vehicles. Those emitting levels of VOCs and nitrogen oxides above standards must be repaired or taken off the road. On occasion, a company will pay to repair cars that flunk the test or, if emissions are bad enough, will buy and scrap the vehicles. Sunoco did this in Philadelphia. The advantage to Sunoco was a partial offset against its own increased emissions when it expanded its refineries.

Changing gasoline composition. Changing gasoline composition is a means to reduce exhaust emissions 30% by the year 2000, especially in locales now not complying with EPA's ozone standard. Oxygen-containing chemicals, such as ethanol, methanol, and methyl tertiary butyl ether (MTBE), are added to gasoline to make it burn more cleanly and reduce carbon monoxide emissions. Cars, trucks, and other vehicles, including tractors and agricultural equipment, can use reformulated gasoline, which costs 3 to 5 cents per gallon more than conventional gasoline. MTBE has proved to be a controversial additive because some claim that its odor makes them ill; potential health effects of MTBE and other oxygenated fuels are being further studied.

Alternative fuels. Instead of just modifying gasoline, *alternatively fueled vehicles* (AFVs) use different and less-polluting fuels, such as natural gas, electricity, propane, and alcohol. In 1992, 350,000 propane-fueled vehicles were in use in the United States. Conversion to propane costs $1,000–2,000 per vehicle, but because propane costs less than gasoline or diesel fuel, fuel savings may pay for conversion costs within a year. Vehicle maintenance costs and engine wear are also reduced because propane is clean-burning and leaves no lead, varnish, or carbon deposits. Another AFV, which will be discussed later, uses hydrogen fuel cells.

An advantage that petroleum has over alternative fuels is that an infrastructure to use gasoline is in place. For AFVs to be successful, separate infrastructures must be available to sell them and their component parts and to fuel them. Specially trained mechanics will also be necessary. To assist in infrastructure development, the federal government plans to replace its gasoline-powered fleet with AFVs. Local governments and private companies will be enlisted in this initiative to help it off to a strong start.

Modifying the vehicle. Automakers are required to introduce low-emission vehicles – vehicles with emissions reduced by 90% – by the turn of the century. One means to do this is to lower *cold-start emissions*, which occur when a vehicle is first started up; these account for up to 75% of all emissions for a typical trip. One proposed means to do this is to develop a hydrocarbon trap. Another vehicle modification is the addition of a device to reduce the vapor emissions that humans are exposed to during refueling; this is being phased into new light trucks and passenger cars.

The first automaker to announce a vehicle with a 90% reduction in emissions was Honda, which modified its Accord EX engine to reduce emissions 90%, as compared to 1994 levels. Manufacturers are also working to develop *zero emission vehicles* (ZEVs), which, by definition, give off no VOCs, nitrogen oxides, or carbon monoxide. ZEVs are often considered synonymous with electric vehicles. However, vehicles powered by hydrogen fuel cells also fulfill the criteria for ZEVs. They emit little or no pollution with less engine noise and lower maintenance costs. There is hope that hydrogen-powered vehicles, at least buses, will be available by the year 2000. As of 1996 Daimler-Benz had developed a prototype car fueled with hydrogen fuel cells. This company also has prototype urban buses and plans to produce them commercially in 5

BOX 12.2

A first generation. By 1997, two manufacturers had electric vehicles (EVs) ready for the California market. One is the General Motors EV1. It will have 1,200 pounds of batteries under its hood – a pack of 26 12-volt lead-acid batteries. Fully charged, they give the EV1 a driving range of 70 to 90 miles, adequate for the average person driving 55 miles per day. The vehicle can be recharged in 3 hours, running off a 220-volt (30-ampere) power source or in 15 hours off a 110/120-volt (10 ampere) source. The price is $33,995. Several other American manufacturers have EVs almost ready for the market, also powered by lead-acid batteries. However, Honda Motor Company is introducing an EV powered by the more advanced nickel metal hydride battery, which can drive 375 miles on a single charge. Its cost is prohibitive at $74,495, but it reportedly will be leased at a competitive rate.

The biggest problem for manufacturers of EVs is finding effective batteries. As an author in a publication of the Institute of Electrical and Electronics Engineers put it, "A bevy of battery makers are mounting furious searches for the Holy Grail – a reasonably priced, long-life battery pack that will be easy to recharge and free of corrosive materials. But like the Holy Grail, this might be beyond the grasp of mere mortals."

years. However, vehicles fueled by hydrogen cannot yet compete economically with those fueled by petroleum, so introduction may not be rapid. If solar-powered vehicles ever become practical, they will also be ZEVs.

California, with its severe air pollution problems, will require that a small, but increasing percentage of cars sold there be ultra-low-emission vehicles starting in the late 1990s (see Box 12.2). Similarly, 11 northeastern states (the ozone transport states, OTC) may require that cars sold there emit 70% fewer pollutants by 1999 as compared to 1995. However, EPA prefers a nationwide plan and, as part of a 1996 agreement with EPA, auto manufacturers will market low-emission vehicles throughout the United States by 2001. Another low-emission vehicle is a hybrid electric vehicle (HEV), which has both an electric motor and a gas engine. Most exhaust gases are emitted during stop-and-go traffic when the engine is idling. The HEV would use electric power at such times and never idle. However, reaching the highway, the gasoline engine would come on. Many see the HEV as a compromise that could be used until better electric vehicles can be produced: electric vehicles that don't use lead-acid batteries, that can be driven longer distances before recharging, and that are more economical.

Despite its problems with motor vehicle emissions, the United States has the most stringent regulations to reduce emissions in the world. Beginning in 1996, European Union (EU) countries started work to lower car emissions there to half those of 1994. This policy will bring EU countries into line with U.S. standards.

Totally redesigning motor vehicles. The Partnership for a New Generation of Vehicles, better known as the Clear Car Initiative, is a 10-year cooperative effort between the Big Three automakers and the federal government. A major DfE effort, it intends to develop totally new vehicle prototypes. Engineers were asked to choose the most promising new technologies by 1997, create and have a concept vehicle available by 2000 and a production prototype by 2004. Specific objectives are to improve fuel efficiency as much as threefold (to 80 mpg) compared to 1993 while reducing emissions and improving vehicle safety. One requirement of such a dramatic increase in fuel efficiency is that cars be 25–30% lighter than they are now. Strong lightweight materials are already available and are used by NASA in space vehicles. However, their cost is too high to consider using them in cars and trucks. So one major research thrust is to develop lightweight materials such as lighter metals, engineered plastics, and ceramics. Actually, a threefold increase in fuel efficiency is not possible using solely internal combustion engines, so new systems, for instance, hydrogen fuel cells, are being intensively examined, as are other possibilities such as the hybrid gasoline-electric vehicle noted earlier.

> ## BOX 12.3
>
> **Back to the individual.** A study of America's transportation problems by Congress's former Office of Technology Assessment (OTA) concluded that we must focus on how individual Americans use cars and light trucks if we want to reduce energy use and traffic congestion. Personal motor vehicles account for more than 50% of the energy this country uses in transportation. The OTA study, like others before it, concluded that we need cars and trucks with better fuel mileage. It also suggested that fees be imposed on drivers to reflect the many driving services we receive from the government, of which the provision of highways is only one.

As of early 1997, the Initiative was criticized in an *Issues in Science and Technology* article by Daniel Sperling, director of the Institute of Transportation at the University of California, Davis. He noted that the deadlines were too ambitious and that an initiative designed to develop revolutionary new technologies was instead working on incremental refinement of known technologies; neither was it focused on greatly reducing exhaust emissions. Dr. Sperling recommended removing the 1997 deadline to select new technologies. This would allow engineers more time to explore, test, and design the most promising technologies while still working toward the goal of creating a production prototype by 2004. He also urged imposing more stringent emissions requirements on the new vehicles.

Another goal of the Clean Car Initiative is vehicle recycling; means are being researched to recycle bumpers, instrument panels, seats, and interior trim. This is a complex task because many materials are involved: plastics, glass, fluids, sealers, fabrics, adhesives, paint, and rubber. A cooperative effort between businesses that manufacture automotive plastics and the plastics industry has arisen with the intent of developing the technology to recycle plastics from discarded vehicles.

DISCUSSION QUESTIONS

1. An advisory commission to President Clinton concluded that the single most important step the United States could take to reduce greenhouse gas emissions from personal motor vehicles would be to raise CAFÉ standards to 45 mpg by the year 2005. They believed this enhanced fuel economy could be attained using current technology and without sacrificing car safety. In practice, because of pressure from manufacturers, Congress is very reluctant to raise CAFÉ standards.

     (a) Why do you believe manufacturers resist an increase in fuel economy?

     (b) Should they be forced to increase CAFÉ standards? Explain.

2. Raising a car's fuel economy puts the focus on manufacturers. But the OTA stressed that, if we want to reduce energy use and traffic congestion, we must focus on how individuals buy and use motor vehicles.

     (a) Is it possible to raise the consciousness of hundreds of millions of individuals on this issue? If so, how?

     (b) What should be the balance between the manufacturer's and the individual's responsibility?

     (c) Are there changes you are willing to make in the way you buy and use motor vehicles? Explain.

POLLUTION FROM ELECTRICITY PRODUCTION AND USE

We use electricity for a multitude of purposes in our everyday lives. Home and commercial lighting, heating and cooling, and household appliances are some of these. Industrial motors use about 50% of United States electric power and electric lighting another 25%. Electronic applications use increasing amounts; computers alone accounted for about 5% of electric energy in 1994. Although electricity generation is a major source of pollution, some, as discussed earlier, see electricity as a means to reduce motor vehicle pollution, and California is mandating that 1% of all vehicles sold there by 1998 be ZEVs and 10% by 2001. Although electric vehicles do not emit pollutants during use, they are fueled by a process that does: electric power generation. Electric car proponents argue that emissions from a limited number of electric power plants are better controlled than emissions from millions of individual vehicles. Electric power plants are also usually outside the most polluted urban areas and thus relieve urban pollution. Not everyone at EPA is convinced of the superiority of electric vehicles; a 1994 EPA report suggested that electric vehicles, all factors considered, may be less energy-efficient than gasoline-powered ones and that power plants would emit more pollution if called upon to produce power for electric vehicles. However, a counterargument is that, if vehicles are recharged at times other than when peak use of electric power is occurring, new power plants may not be necessary. EPA, with some misgivings, gave the go-ahead to California to require that a small percentage of vehicles sold there be electric. In 1994, EPA's *Pollution Prevention News* referred to electric vehicle use as P^2, but whether electric cars are truly a net environmental plus is not settled.

What are the pollutants of concern from electric-power generation?

The pollutants produced by electric-power generation can be deduced from the way electricity is produced: Seventy percent is generated from burning fossil fuels, especially coal (see Figure 12.2). Worldwide, the figure is similar. Pollutants produced by each of the fossil fuels are considered next followed by a discussion of alternate means of generating electricity.

Coal

- Air pollution: Coal has a higher carbon content than petroleum or natural gas. Thus, burning coal produces more of the greenhouse gas carbon dioxide than burning other fossil fuels. Coal-burning power plants are also the major source of sulfur dioxide, which is a major precursor of acid rain; they

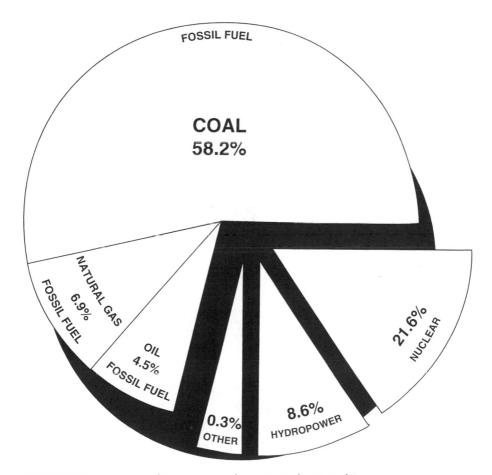

FIGURE 12.2. Sources used to generate electricity in the United States.

also emit nitrogen oxides (another precursor of acid rain), particulates, radionuclides, and metals (see Table 12.1). Coal contains only trace amounts of metals, but, because massive amounts of coal are burned, emissions are significant.

- Water pollution: Table 12.1 shows only the air pollutants that are formed by burning fossil fuels. Burning coal also contributes to water pollution through runoff from mining operations and ash disposal sites. In addition, many air pollutants emitted when coal is burned settle into water.
- Soil pollution: Air pollutants from coal burning settle onto soil, and huge quantities of ash must be disposed of on land. The ash contains higher concentrations of metals than the coal itself because the organic material has been burned away leaving inorganic substances such as metal oxides.
- Other issues: Strip mining is tremendously destructive to land, and underground mining is often unsafe to miners. Despite problems, coal will doubtless continue to be used if for no other reason than that tremendous coal reserves exist in the world, about half of them in the United States. The challenge is to develop technologies that allow us to burn it with less pollution and to mine it in a way that respects the land and human life.

Petroleum. Petroleum burns more cleanly than coal. Nonetheless, all the air pollutants produced by burning coal are also produced by burning oil, ordinarily in a less severe form. In addition, marine oil spills continue to occur, sometimes with devastating effects. Land transport sometimes leads to fires, explosions, and loss of life. Although petroleum is still plentiful, it is a nonrenewable fuel. The United States increasingly depends on foreign sources, and an estimated 80% of our oil will be imported by the year 2030. Oil will not run out in the foreseeable future, but its use will become more costly as easy-to-tap sources become depleted, forcing the use of those more difficult to tap. The real challenge is to find means to reduce American reliance on oil.

Natural gas. Compared to other fossil fuels, natural gas burns most cleanly and produces the least carbon dioxide. But because natural gas is primarily methane, gas pipe and other leaks of natural gas increase the atmospheric concentration of this greenhouse gas, which is about 20 times more potent than carbon dioxide. Also, natural gas use periodically results in explosions, which are especially destructive when they occur in an urban setting. Another safety concern is the hundreds of carbon monoxide deaths occurring each year in private homes as a result of malfunctioning natural-gas furnaces. Natural gas is used in increasing amounts, is plentiful, and is expected to stay so well into the next century. But, like other fossil fuels, it is nonrenewable.

How is the pollution associated with electric power generation being reduced?

Conservation. The ideal P² approach is conservation. Use less electricity in homes, institutions, and industry. As noted in Chapter 6, there are a number of ongoing efforts to reduce industrial use of electric energy, especially of the motors that consume 50% of United States electricity. There are also many ways that an individual can reduce energy consumption. Begin by becoming aware of how you use energy. Check to find places in your home where energy-efficient light bulbs can be used. These fit into regular sockets, but don't accept a traditional shade. Don't use lighting unnecessarily: Turn lights off when you leave a room; turn the thermostat down at night; look for and buy appliances that use energy efficiently. Refrigerators and clothing dryers are particularly heavy electricity users. Insulate and weatherproof your home, while taking care to maintain good ventilation.

Changes in processing fossil fuels. Better processing can be P². For example, clean coal technologies remove a portion of coal's sulfur, allowing it to burn with lower sulfur dioxide emissions. Ideally, clean coal technology needs to remove some of the metals whose emissions also contribute to the environmental damage of coal burning. Sulfur can also be removed from petroleum before it is burned.

In another P² approach, coal is burned on fluidized beds. In this technique, coal is pulverized and burned with ground limestone on a screen. As particles burn, an air current keeps them in constant motion, giving the appearance of a boiling or fluidized bed. The ground limestone with which it is burned captures the sulfur dioxide generated. The fluidized beds allow more efficient burning, and thus fewer products of incomplete combustion are formed and emitted.

Burning of less polluting fossil fuels. Natural gas is the cleanest fossil fuel, petroleum is intermediate, and coal the dirtiest. Preferentially burning natural gas is P², generating fewer of almost all the pollutants noted in Table 12.1. This includes lowered emissions of carbon dioxide because natural gas contains the smallest percentage of carbon. Preferentially burning naturally cleaner coals is also P². High-sulfur coal contains up to 3% sulfur by weight and produces large amounts of sulfur dioxide when burned. However, some western U.S. coals have sulfur contents as low as 0.3% and, when burned, have much lower emissions of sulfur dioxide.

Switching to renewable fuels. Convert the energy of wind, sunlight, falling water, biomass, or geothermal energy into electricity. As renewable resources, these are considered superior to fossil fuels, but none is without environmental problems, as will be discussed. Worldwide use of renewable energy sources is expected to more than triple by the year 2020, but, as a percentage of total energy used, the picture is disappointing. This is true because, although use of alternative energy sources is increasing, so are population and industrialization. These will lead to an anticipated increase in energy demand of 30% by 2025. Another problem is the expense of building the infrastructure necessary to produce and use new energy sources. A 1993 report from the World Energy Council notes that, even if renewable sources are aggressively promoted, they will furnish only 30% of the world's energy in the year 2020 with fossil fuels and nuclear power still accounting for 70%. More positively, the Council predicted that carbon dioxide emission should have fallen 25% by 2020, as compared to that in the early 1990s. Two-thirds of the decrease would be from advances in energy efficiency and only one-third from the use of renewable energy.

Others are more optimistic than the World Energy Council about the future of renewable energy sources. According to one scenario developed by Shell Oil Company, renewable energy technologies may be competitive with fossil fuels by the year 2020. Looking yet further ahead, Shell analysts do not believe that much coal, oil, and gas will be used 100 years from now. Further, they optimistically project a sustainable energy future. A *Scientific American* author, reviewing the Shell Oil study in November 1994, stated, "When a major oil company effectively projects the end of the fossil fuel age, it is a sure harbinger that we are moving into the future on fast-forward."

RENEWABLE ENERGY SOURCES

This section examines some of the advantages and problems associated with the use of alternate energy sources. As will be seen, P^2 is not an easily applicable concept when considering these. This is true because an alternative may produce few or no pollutants and use fewer resources, but have other adverse impacts.

Hydroelectric

In 1990, hydroelectric dams produced electricity equivalent to a half billion tons of oil. By the year 2040, this figure is expected to rise to the equivalent of 1.4 billion tons. A dam appears to have many environmental advantages. Unlike fossil fuel, it generates no air pollution and does not deplete nonrenewable resources. But it may flood land; deprive downstream wildlife, farmers, and recreational users of water; interfere with fish migration; and otherwise disrupt downstream ecosystems. It also has health and safety implications. Beyond these issues, problems associated with water use can cause angry confrontations. Even for a river flowing within the borders of only one country, angry confrontations can arise among competing users of the water. For international rivers, dams can cause severe tensions. They also have a limited lifetime because of buildup of silt.

Solar

Every year sunlight provides the earth's surface with an estimated 10 times as much energy as is stored in all known reserves of coal, oil, natural gas and uranium combined. Humans tap into solar energy when they burn wood and other biomass. The sun's energy is also stored in other ways. Fossil fuels originated from biomass. Energy resulting from the flow of water over dams likewise has its origin in the sun's energy, which serves to evaporate the water that subsequently falls onto the earth as precipitation. The energy of the wind that drives wind machines also comes from the sun.

The term *solar energy* as applied to electric power generation, typically refers to photovoltaic cells or thermal solar systems.

- Photovoltaic cells directly generate electricity from light, which can be used for such small scale applications as powering a calculator or, sometimes, a home's appliances.
- A thermal-solar system produces electricity indirectly. Sunlight is reflected from panels onto collectors; in the collectors, water (or sometimes other substances) is heated to produce the steam used to power the turbines that make electricity. Thermal-solar energy can also be used to bypass electricity generation entirely. An example is its use to heat a home's water supply as water is pumped through roof panels.

The use of solar power does not directly pollute, but pollution is generated when the solar panels are manufactured, transported, and maintained, and workers must be protected from the toxic chemicals used to manufacture photovoltaic cells. To produce a significant amount of electric energy, a large area of land must be covered with panels. Still, the U.S. Electric Power Research Institute estimates that a quarter of America's electricity needs could be met by less than 6,000 square miles of solar farms, an area equivalent to the state of Connecticut. The electricity from sun-powered plants must often be transported longer distances than from local power plants fueled with fossil fuel. This transportation may lead to confrontations with those who do not want power lines running across or near their lands or close to homes or schools.

As part of making solar energy practical and economical, technology must be developed to store the energy so that it can be used at night and on days with little sunlight. One possible storage mode is to convert solar to chemical energy; the "energy-rich" chemicals formed could then be transported to the sites where fuel is needed to generate electric power. To prevent problems associated with transporting solar power, some homeowners use solar panels to generate their own solar power. This becomes progressively more difficult the farther north one lives and in cloudy climates.

Wind machines

Like solar-energy devices, wind machines do not directly pollute, but rather generate pollution as they are manufactured and transported and when access roads are built. At least some bird deaths result from wind turbines, and noise pollution may be a problem. Large tracts of land are needed, but, except for the actual surface covered by machines and access roads, the land can be used for crops or trees. To be effective, there must be dependable wind. As with solar power, there must be means to store electric power, in this case, for periods when the wind does not blow. As with solar and hydroelectric power, transporting electricity from wind machine sites to customers can present problems. California wind machine farms have the potential to furnish 5% of that state's electricity, although, because wind speed is seldom optimal, they currently produce only about 1%. Wind farms could potentially provide a quarter of American electricity needs.

If 25% of U.S. electricity needs were generated from solar and 25% from wind power, this would represent a significant change from the present situation, in which 70% of electricity is furnished by fossil fuels. As with solar power, there are small-scale uses of wind machines; they were once widely used in the Midwest to pump water. Modern wind machines are now sometimes used to generate electricity in private homes.

Biomass

Biomass fuels include wood and agricultural wastes. If grown in a sustainable manner, biomass has no net carbon dioxide emissions because carbon dioxide released during biomass burning equals carbon dioxide absorbed as new biomass grows. However, it is difficult to burn biomass cleanly and some types are available only seasonally. Furthermore, the land used to grow biomass may be more valuable for other purposes; consider an increasing population and the corresponding need for agricultural land. However, land not suitable for growing food may support certain biomass crops. Denmark is one developed country that is making effective use of biomass (see Box 12.4). A challenge to

third world countries using biomass is the need to find relatively nonpollut-
ing, easily grown biomass sources to fuel cooking stoves, eliminating the need
to generate electric power for cooking uses.

Municipal solid waste

MSW can be burned to generate electricity and is always available in what
seem to be very large amounts. However, when compared to electricity needs,
MSW is only a small source. Also, in some cases, MSW that is burned might be
better recycled. Most scientists believe that with safeguards and vigilance we
can burn MSW cleanly and landfill ash safely, but concerns remain. If we continue to combust MSW, means must be found to motivate individuals, communities, and businesses to remove batteries, electronic components and other
metal-containing materials from their trash. This allows MSW to be more
safely burned while also allowing the metals to be reused. Finding alternate
environmentally safe uses for the ash generated is another challenge.

Geothermal energy

In locales like Iceland and California where it is available, geothermal energy
is relatively cheap. However, it can create local air pollution from chemicals

---

BOX 12.4

Straw and electricity. In contrast to many other countries the United
States keeps oil prices kept artificially low (see Table 12.2). Keeping
prices low means we use more oil than we would otherwise and have less-
ened incentive to conserve energy or to pursue alternate energy sources.
Denmark is taking a different route. At the time of the Arab oil embargo
in 1973, 90% of Denmark's energy needs were met by imported oil; the
embargo shocked this small country into developing a different energy
policy. Energy-conserving measures were diligently taken and alternate
energy sources developed. By 1988, its dependence on imported oil had
declined to less than 50%, and it continues to decline. Biomass already
supplies 6% of Denmark's energy consumption, and the country is now
converting some coal fired electric power plants to biomass. By the year
2000, Denmark projects that 7%-8% of its electricity will be generated
from burning straw with some input of wood. Burning straw presents
many technical problems, but the effort is expected to be successful. In
addition to reducing oil dependence, straw is renewable and, grown in a
sustainable manner, will not increase atmospheric carbon dioxide.

TABLE 12.2 Price of a
gallon of gasoline
(in U.S. dollars)

COUNTRY	PRICE
Venezuela	0.44
Ecuador	1.12
U.S. (New York)	1.36
Russia (Moscow)	1.89
Nigeria	1.89
Brazil	2.46
Israel	3.29
Argentina	3.50
Tokyo	3.95
Sweden	4.43
France	4.50
Norway	5.03

Source: Christian Science
Monitor, May 24, 1996.

drawn to the surface with the steam. Local geothermal supplies can also be
overdrawn.

Hydrogen

Hydrogen is a gas that can react with oxygen to form water; this reaction produces energy with almost no pollution. For example, unlike fossil fuel burning it produces no VOCs and emits no metallic contaminants, nor (because it
contains no carbon) does it emit the greenhouse gas carbon dioxide. The
problem has been to generate hydrogen economically and in a form convenient and safe to use. Hydrogen can be obtained by electrolysis, a process in
which an electric current is run through water (H_2O) to produce hydrogen
(H_2) and oxygen (O_2) – the reverse of the reaction noted in this paragraph's
first sentence – *but* it takes energy to generate the electricity that electrolyzes
the water; thus there cannot be energy gain unless efficient means are found
to fuel the process. Such means are being explored; for example, solar thermal cells or photovoltaic cells can be used to generate the electricity to elec-

trolyze water. Or the hydrogen can be generated by means other than the electrolysis of water. To stimulate the research and development needed to make hydrogen fuel a practical reality, the U.S. Congress passed the Hydrogen Future Act in 1996, which provides funding for this purpose. As noted, hydrogen already fuels a few vehicles and has long-term potential to fuel electric power plants. Although it can be transported long distances, it is explosive and flammable, and the necessity to store it in pressurized tanks increases its danger and makes it more difficult to use. Better technology is needed to address safety concerns; one option is to convert it to solid forms (hydrides, for example) that are easier to use than gaseous hydrogen and more cost-competitive.

Nuclear fission

No air pollution is produced by nuclear power plants. Although some components of nuclear waste are very long-lived and some are highly radioactive, waste volume is very small. There are two major objections to the use of nuclear power: One is a concern that nuclear reactors cannot be operated safely; the other is disposal of nuclear waste. Both issues are tremendously controversial. American nuclear reactors are different from and much safer than those involved in the 1986 Chernobyl accident in the Ukraine, but, safety concerns remain. Some believe that radioactive wastes cannot be safely disposed. Others believe the problems are political, not scientific. Nonetheless, no new nuclear power plants are being built or being planned in the United States. Nuclear fission may have been a much less controversial energy source if safety and radioactive waste concerns had been handled early in the history of its use. However, the initial user of nuclear technology was the military, who did not consider health, safety, and environmental concerns a priority.

Nuclear power proponents make the following comparison between a 1,000-megawatt nuclear power plant and a 1,000-megawatt coal-burning power plant: The coal-burning plant emits 35,000 tons of sulfur dioxide and 4.5 million tons of carbon dioxide, and produces 3.5 million cubic feet of ash a year. It also emits metal pollutants, which are present in coal only in parts per million. But the sheer amount of coal burned – 616 million tons in 1982 – leads to large total emissions. Radioisotopes are also emitted at levels well above those that are legal for a nuclear reactor. In 1982, coal burning emitted 801 tons of uranium and 1,971 tons of thorium into the U.S. environment. In contrast, the 1,000-megawatt nuclear plant produces 70 cubic feet of high-level radioactive waste. Such waste can be vitrified (embedded in a stable glass) and stored in a deep underground repository in a stable geologic formation. However, burying nuclear waste is so controversial that no permanent site is yet available in the United States to do this.

Nuclear proponents believe that plants with advanced reactor designs, that take the performance of imperfect human operators into account can be operated safely. However, even if no new nuclear reactors are built, we must deal with the nuclear waste that already exists. Waste from commercial reactors and yet smaller amounts of wastes from medical and research uses represent small quantities as compared to the nuclear wastes found at sites formerly operated for military uses. These sites represent a costly and technologically challenging cleanup problem. One positive step is an active program investigating new approaches to treat nuclear waste that will greatly reduce its volume and the life span of its radioactive chemicals. Success in this effort could help the civilian nuclear power industry. Nuclear energy still fills about 20% of U.S. electric power needs and plays an even larger role in France and Japan. France vitrifies its nuclear waste. In this process, the waste is embedded in impenetrable insoluble glass. There is skepticism as to whether the glass can be trusted for the many thousands of years required for the longest-lived radioactive components to decay; however, nuclear proponents point to examples such as the Oklo "natural reactor" in the Republic of Gabon; starting about 2 billion years ago, natural uranium fission reactions took place at this site over a 500,000-year period, and hundreds of tons of nuclear fission products formed. Over a 2-billion-year period, none of those radioactive products has migrated from the rocks. Critics point out that there is no assurance that other storage sites would be equally safe.

Nuclear fusion

Unlike nuclear *fission*, nuclear *fusion* does not use radioactive chemicals. The process does produce neutrons, and it is the capture of these neutrons that produces heat. The reaction vessel becomes radioactive during this process but remains radioactive for a much shorter time than do isotopes produced by nuclear fission. Unlike fossil fuel burning, nuclear fusion produces no air pollution. Research and development for this technology have been extremely expensive and have proceeded very slowly. Congress is funding a joint research program with the European Union, Japan, and Russia intended to demonstrate self-sustaining fusion by 1998 and start operating a reactor by 2005.Whether this prospect is likely is not clear. Although many scientists consider the likelihood of success dim, the Japanese are funding research into an alternate mode of nuclear fusion, so-called cold fusion.

Other energy sources

There are many other potential energy sources. The ocean's tidal energy could be harnessed, though some fear that deriving power from tides would disrupt whole ecosystems. Others believe such environmental problems could be overcome. Another means to garner energy from the oceans, believed to be more environmentally sound, is ocean thermal energy conversion.

DISCUSSION QUESTIONS

1. It is 1945. World War II is over, ended after two nuclear bombs were dropped on Japan. You strongly desire to see nuclear energy used for peaceful purposes and are especially interested in nuclear electric power. However, you find yourself thinking through problems that would be associated with nuclear power. You know that ionizing radiation can be very harmful to life, understand that nuclear reactors are not always safe, and realize that there are limits to fail-safe operation: Even if reactors are safe, human operators are imperfect. You also know that some components of nuclear waste remain radioactive for many thousands of years and that disposal of them poses perplexing problems. Nuclear power technology can also lead to development of nuclear weapons in countries that do not now have them.
   (a) Do you still want to see nuclear power developed?
   (b) If so, what steps should society take to deal with these problems in an environmentally and socially responsible way?
   (c) If you do not want to see it developed, find arguments to persuade your colleagues.

2. Do you believe Americans will develop confidence in nuclear power again within your lifetime? Explain.

3. Compare the environmental impacts of coal burning to those of an hydroelectric dam.
   (a) What impacts of each most concern you?
   (b) Can the two be directly compared?
   (c) If not, how can society make comparisons of this nature and thus make decisions on how best to produce energy?

4. (a) Among the alternatives to the use of fossil fuels to generate electricity, which do you believe to be the most feasible? Explain.
   (b) Are these also the most environmentally benign in your opinion?

5. The development of alternatives means of generating electric power is happening very slowly.
   (a) How does individual behavior contribute to that slowness?
   (b) List 10 possible steps to conserve electric energy.
   (c) Which do you already practice?
   (d) What obstacles stand in the way of doing more?

FURTHER READING

Brookins, D. G. 1990. *Mineral and Energy Resources.* Columbus, Ohio: Merrill.

Dreyfus, D. A., and Ashby, A. B. 1990. Fueling Our Global Future. *Environment,* 32(4), 17–20, 36–41, May.

Flavin, C. 1996. Power Shock: The Next Energy Revolution. *World Watch,* 9(1), 10–21, Jan./Feb.

Flavin, C. 1993. Jump Start, The New Automotive Revolution. *World Watch,* 6(4), 27–33, July/Aug.

Hollander, J. M. (ed.) 1992. *The Energy-Environment Connection.* Washington, D.C.: Island Press.

Lee, W. S. 1990. Energy for Our Globe's People. *Environment,* 32(7), 12–5, 33–5, Sept.

Rogers, P. 1991. Energy Use in the Developing World: A Crisis of Rising Expectations. *Environmental Science & Technology,* 25(4), 580–3, April.

Rose, J. 1993. How Much Can Renewable Energy Deliver? *Environmental Science & Technology,* 27(12), 2267, Nov.

Sperling, D. 1996–97. Rethinking the Car of the Future. *Issues in Science and Technology,* XIII(2), 29–34, Winter.

Tunali, O. 1995. A Billion Cars: The Road Ahead. *World Watch,* 9(1), 24–33, Jan./Feb.

United States DOE. *Tips for Energy Savers.* DOE/CE-0143, Washington, D. C.: U. S. Department of Energy.

13

POLLUTION AT HOME

Mammoth environmental problems confront humanity worldwide. In the United States, despite advances, major problems remain. But beyond large-scale problems are those that affect us individually. They are also significant. EPA's Science Advisory Board and others believe that pollution in the home is a significant environmental health risk. This chapter will first overview hazardous household products and then look at indoor air pollution.

HAZARDOUS HOUSEHOLD CHEMICALS

About 1.5 million Americans visit hospital emergency rooms each year because of poisonings or possible poisonings; about 10% are admitted into the hospital. More than 80% of poisoning cases occur in children less than 6 years of age. In addition to emergency room visits, many people have reactions to hazardous household products such as pesticides, solvents, ammonia, or chlorine that they do not report. Exposures to household products are also responsible for injuries to the skin and eyes.

Why are household chemicals of concern?

The words *toxic, corrosive, flammable,* and *reactive* were introduced in Chapter 8 to describe characteristics of industrial hazardous waste. They are discussed here from the perspective of household products.

- A corrosive product can directly damage (corrode) the skin, eyes, or mouth. Examples are products containing alkali (lye) such as drain openers and many oven cleaners. Corrosive products often have the word *poison* on the label and the word, *danger*. In 1988, more than 2,000 people were admitted to emergency rooms for treatment of skin or eye injuries resulting from the use of corrosive drain openers; highly acidic products such as some tile and toilet cleaners are also corrosive. A corrosive diluted with water may only be an irritant and a very dilute corrosive may present no hazard at all. An irritant is ordinarily less dangerous. It can cause skin redness, itching, or rashes; cleaners, cosmetics, metals and polishes are irritants to some people. It becomes more than an irritant when a person becomes sensitized, that is, becomes allergic to it. Some individuals have allergies to metals, chemicals in cosmetics, or latex gloves. These reactions are sometimes severe.

- Household chemicals can be toxic. Because all substances are toxic in high doses, all products, including vitamins, minerals, aspirin, and laxatives, should be treated with care, especially if children are in the household. However, only a few household products fit the legal definition of poison and these are clearly labeled. Among these are oxalic acid, found in a few cleaning products, and methanol, found in windshield washing fluids. Most pesticides available to householders are not as toxic as those available to professional applicators and are not legally considered poisons. Nonetheless, home pesticide poisonings occur. Toxicity information on the label refers to acute toxicity, an adverse health effect occurring soon after exposure. The label does not – often cannot – provide information on possible chronic toxicity.

- A product may be flammable, that is, catch on fire easily. Household products that contain petroleum distillates are flammable; examples are oil-based paints, paint thinners, paint strippers, certain furniture and floor polishes, and some rug cleaners. Products that contain organic solvents are often flammable, as are many aerosol sprays that use flammable hydrocarbons as propellants. A substance that is *combustible* is similar to one that is *flammable* except that it does not catch on fire as readily as a flammable one.

- A product may be reactive. Certain dry and liquid bleaches that contain chlorine will react with household ammonia to give off toxic fumes. They also react with products containing acid as some toilet and tile cleaners do. Even some dish-washing detergents contain ammonia and should not be mixed with chlorine-containing products. To be safe, do not mix chemical products. Overheated or accidentally punctured aerosol cans can be reactive in a different way: They may explode. Well-known reactive substances are dynamite, gun powder, and firecrackers.

- A radioactive product *could* be hazardous but, except for tiny amounts in some smoke detectors, exit signs, and Coleman lamp mantles, radioactive chemicals are not found in consumer products. TV sets emit small amounts of ionizing radiation, as do building materials that come from the earth including granite, other rock, and soil. These have higher background levels of ionizing radiation than does wood.

- A product may have more than one hazard. Gasoline has several: It is flammable and its fumes are toxic. Young people who inhale gasoline fumes for a "high" continue to be poisoned or killed by this practice. A third hazard of gasoline is when people accidentally aspirate it into their mouth when transferring gasoline from a motor vehicle's fuel tank. Even small amounts in the lungs can be disabling or deadly. These same hazards are present in other products that contain petroleum distillate or organic solvents.

Labeling laws. In the early 1900s, about 500 American children per year, mostly under 5 died from accidental home poisonings. Thousands more were burned by the lye (caustic) used to make household soap. This situation began to change in 1927, when Congress passed the Caustic Poison Act requiring that containers of caustic be labeled. In 1960, a law applying to all hazardous products sold to householders was passed, the Hazardous Substances Labeling Act. Finally, the 1970 Poison Prevention Act, requiring child-proof containers for certain hazardous products, was passed. In the 1990s, although the U.S. population has greatly increased as compared to that 90 years ago, many fewer children die from poisoning.

Products with one or more of the hazards noted are found in grocery stores, hardware stores, discount and craft and hobby stores, and agricultural products and drug stores. In short, almost any store may sell products containing hazardous ingredients. The Hazardous Substances Labeling Act established labeling requirements for all consumer products containing hazardous substances except pesticides; their labeling requirements were established under the Federal Insecticide Fungicide and Rodenticide Act (FIFRA). Information that must be on the label of hazardous products is as follows:

- Signal words: *poison, danger, warning,* or *caution.*
- Principal hazard: flammable, harmful or fatal if swallowed, skin and eye irritant, or vapor harmful.
- Common name or chemical name for the hazardous ingredients.
- Name and address of the product's manufacturer or distributor.
- Statement, Keep Out of Reach of Children, or equivalent.
- Precautionary measures: instructions for safe use of the product.
- First-aid instructions: Do not assume these are adequate. Call the Poison Control Center if an accident occurs.

Using labels to detect hazardous products

- The signal words on the label (*poison, danger, warning, caution*) indicate a product's hazards. Look especially for the word *poison* or *danger; poison* indicates the greatest level of toxicity. The signal word *danger* may refer to a corrosive, flammable, or reactive substance.

- For pesticide products, the signal words are slightly different: *Warning* refers to medium toxicity, indicating a product less toxic than one with the signal word *poison*, but more toxic than one with the signal word *caution*.
- *Warning* and *caution* have a less precise meaning in other household products than in pesticides. However, *caution* usually refers to the lesser hazard; detergents are household products that commonly have only the signal word *caution* on the label. Still, home poisoning accidents often occur with detergents; small children are especially likely to ingest them. Fortunately, their taste, combined with low toxicity, prevents most incidents involving ingestion of a detergent from being serious.

The term *nontoxic* is sometimes found on products. Unlike the signal words described previously, nontoxic has no legal meaning; any substance can be toxic in high enough doses. However, a product labeled nontoxic can be assumed to be less toxic; otherwise, legally a signal word would have to be displayed on the label. Some terms found on labels are generic. *Petroleum distillates* is one of these; petroleum distillates contain chemicals that are produced during petroleum cracking. The term *organic solvent* is also generic and may refer to acetone, isopropyl alcohol, methylene chloride, or another solvent. A product's active ingredients are those that carry out the function that the product is designed to do. The inert ingredients in a product carry its active ingredient or make it easy to apply, but some inert chemicals are also hazardous.

How can exposure to hazardous products be reduced?

The first steps in self-protection are to read labels and follow directions. However, more general knowledge is important. In particular, it is important to minimize use of especially hazardous products and to handle those that are used in a way that minimizes exposure.

Toxics use reduction (TUR). TUR is reducing exposure to hazardous substances by eliminating or reducing their use; it is a form of P^2. TUR concepts are most often applied to the work place as a means to reduce worker exposure to hazardous substances, but TUR is also important in the home and office. The first TUR rule is, do not use the hazardous product if it is not necessary. The second is, if the product is needed, use the minimum amount possible and minimize exposure as much as practicable. An example is drain cleaners; because most commercial drain openers are corrosive, avoid them by maintaining home drains with boiling water and baking soda. Use a mechanical snake if clogging does occur. Another example is mothballs, which contain insecticide. To avoid using mothballs, store fabrics clean, remembering that moths are attracted to stains on fabrics.

Avoid products with the signal word *danger* or *poison*; alternatives to such products are often available. Glue: If children will be using it, buy glue with the word *nontoxic* on the label. Paint: When possible, use water-based (latex) rather than oil-based paint. Furniture polish: A little lemon juice in vegetable oil is a safe alternative to a commercial product containing petroleum distillates. Oven cleaners: Practice preventive maintenance by keeping the oven clean with baking soda and abrasive scrubbing materials. If purchasing an oven cleaner, try one of the new noncorrosive cleaners.

Using hazardous products. When hazardous products are used, read and follow directions on the label. Always use plentiful ventilation. Develop the habit of using gloves and safety glasses. Do not mix chemical products; in particular, never mix bleach and ammonia. Keep chemical products away from children. This latter precaution includes detergents and vitamin and mineral supplements as well as products more ordinarily thought of as dangerous. Always keep products in their original containers; this is especially critical to prevent poisoning children. Do not use aerosol products in closed rooms. Even when using latex rather than oil-based paint, provide plenty of fresh air; use an exhaust fan to remove the fumes from volatile petroleum-based products such as paints or shellacs. At the very least, take regular fresh air breaks and never use these products in a closed room. If paint must be stripped from a piece of furniture, wait until summer and do it outside.

Hazards dating from earlier years

A number of hazards are unique to older homes. Prominent among these are asbestos and lead paint.

Asbestos. Asbestos is a natural fibrous mineral material. Because of its heat resistance, asbestos was widely used to insulate furnaces and furnace pipes, and it is still found in some old homes and buildings. Such asbestos presents the greatest danger because it may be flaking, allowing particles to be airborne. It is worthwhile to know that asbestos was also a component of floor and ceiling tiles, fireplace gloves, and ironing board covers, which are less likely to present problems. Exposure to asbestos, especially at levels found in industrial settings, is associated with asbestosis, lung cancer, and mesothelioma. Some risk is associated with ingesting asbestos in drinking water, but the primary risk arises from inhaling airborne fibers into the lungs, where they may become permanently trapped. Within the home, an amateur cannot tell whether asbestos is present in old tiles or other materials. Thus, buyers of older homes may want to have them examined by a professional trained to identify asbestos. Fortunately, chrysotile asbestos (considered less dangerous than amphibole asbestos) was used in most buildings and homes. Removing asbestos from heating systems is not recommended unless it is deteriorating because the removal process releases asbestos fibers into the air. Instead, find out from a professional how to maintain it in place. Likewise, it is recommended that home tiles containing asbestos be left in place unless they are deteriorating. Repair or removal of asbestos-containing materials should be done by an EPA-approved professional.

Lead. Lead water pipes have not been used for many years, and lead solder in plumbing was banned in 1988. But, because many people have homes built prior to 1988, lead hazards remain. Lead leaches into water most easily when the pipe or solder is new. Thereafter, mineral deposits settle inside the pipe and less lead is leached from it. The highest lead levels are found in tap water run the first thing in the morning after the water has been in contact with the pipes for many hours, long enough to dissolve lead into the water. Running the tap until the water begins to run cold before using it for cooking or drinking can eliminate much of this lead. Because lead dissolves more easily in hot water than in cold, use water from the cold water tap for cooking. These precautions are not necessary if you know that no lead-containing pipes or solder is in the plumbing system. Call your municipal water system to obtain specific information on testing your pipes for lead. Health concerns associated with lead were discussed in Chapter 9.

Lead was added to paints for many years although in progressively lower amounts until it was banned completely from indoor household paint in the 1970s. Paint with high levels of lead is most often found in homes built before 1940, but paint with lesser amounts can be found in homes built through the mid-1970s. Local and state health departments can provide information on identifying lead paint. Tiny children are most susceptible to lead's adverse effects, and their exposure poses the most concern. They may eat flaking lead paint from window sills or woodwork. A more important route of lead exposure for most children is inhalation of airborne dust in homes that contain leaded paint. Vacuuming the home may simply exhaust the lead-containing particles back into household air. Unless an especially equipped vacuum cleaner is available, a better method is to use wet cloths and mops. As with asbestos, if the lead paint is in good condition, it is best to leave it undisturbed. Wall paper may be used to cover walls with lead paint. If painted woodwork is in good condition, wipe it frequently with a wet cloth to lessen the chance that lead dust will enter the air. If small children are in the home, consider having the paint removed from woodwork by a professional or replace the woodwork entirely. Homeowners should not attempt to remove leaded paint themselves. Soil around old houses often contains high lead levels, which can be tracked into the home on footwear.

Reducing household hazardous waste

Residues of hazardous products become household hazardous waste (HHW). Creation of HHW can be prevented by buying the smallest amount of a product that can do the job desired. This is an instance when economy sizes should be avoided. Once a hazardous product has been purchased, use it up or give it to someone who will. If residues of the product remain, dispose of them carefully; for example, dry out waste paint in a place safely removed from children before disposing of it in the trash. Sometimes product labels have disposal instructions or a telephone number to call for more information on the product.

Used oil and antifreeze. Used oil should never be poured on the ground. In particular, it should never be poured down a drain in the home nor into a storm drain. Not only is the oil itself a pollutant, but it has picked up metals while circulating through the vehicle's engine. Check your city office to find locations that handle used oil. Although it may be difficult to find facilities that take used antifreeze, it also should never be poured on the ground or the driveway. It is sweet tasting and can poison animals and children; some brands now have a bitter-tasting additive to discourage its ingestion. Contact the local wastewater treatment plant to inquire whether they can handle antifreeze. If it can, flush the antifreeze down the toilet, little by little. Do not pour antifreeze down the drain if the home has a septic system, as it may kill the system's microorganisms.

Paint. About half of all household hazardous waste is paint. When paint is being used, it is the number one contributor to indoor air pollution. Oil-based paints containing mineral spirits or petroleum distillates contribute most heavily to indoor pollution and are also toxic and flammable. Water-based (latex) paint is less hazardous than oil-based paint. "Natural" paints that contain beeswax, plant waxes, and linseed oil are available, but check their labels for expiration dates. Respect those dates because some can, quite literally, rot. Buy only the amount of paint needed. If leftover paint is usable, give it away. Encourage your community and state to become more active in a paint-recycling program or in a paint "drop and swap."

Pesticides. Pesticides are used in about 95% of American households. The products most used by homeowners are disinfectants, moth repellents, pet flea collars, no-pest strips, and lawn and garden pesticides. The pesticides available for household use are ordinarily less hazardous than commercial pesticides, and homeowners use smaller amounts, but basic problems remain. They may be blown by wind or carried off with rainwater into other people's yards or into a local stream or lake. They may kill beneficial insects such as honeybees. Because they are picked up on footwear and tracked indoors, pesticide residues are detected on carpets even when the homeowner has not used pesticides. The person applying the pesticide usually gets the most exposure, but family members and others using the home and yard are also exposed.

- Pesticide use within the home can be reduced by P². Keep the home free of food particles, which attract roaches and ants, and store food in tight-fitting containers. Fill in cracks that could allow roaches and other pests to enter the home. A fly swatter can often replace an insecticide. "Natural" repellents are sometimes useful. Some people report that cedar chips in their dog's bed prevents fleas or that pyrethrum daisy flowers help deter moths. Boric acid is useful in cockroach and ant control and is less toxic than many other pesticides. To prevent attracting moths, store clothing clean. In a home with carpets or small children, leave shoes at the door to prevent tracking pesticides into the home.
- Outside the home, consider allowing dandelions, clover, and other "weeds" to be part of the lawn rather than using a pesticide. In the garden, find means of using P² instead of pesticides by reading up on companion planting as a means to help deter predators and on rotation of crops from year to year to lower pest populations. Reduce places, such as spots with free-standing water, that can allow pests to multiply easily.
- When a pesticide is needed, buy the smallest amount that will do the job because residues become HHW. If a spray is used, buy a pump, not an aerosol; the latter produces tiny aerosol particles, which can be breathed deeply into the lungs. Read the pesticide label and follow directions. If using a pesticide within the home, make sure plenty of fresh air is available. Check if gloves are needed. Check for how long a period after pesticide application a treated area should be avoided, and, in particular, think about steps to protect children using the area. Dr. Michael Shelby of the National Institute of Environmental Health Sciences has made the following recommendation on pesticide use: "The best advice for anyone, and my advice at home for my wife and children, is (1) don't overuse them; (2) don't expose yourself to them, or if you do, do it at absolute minimal levels; and (3) keep them in a safe place where animals and children cannot get to them."

HHW collection programs

In the hands of a householder, HHW is not regulated as hazardous waste. However, once it is turned over to a collection program, it is. Thus, community HHW collection programs are expensive and sometimes a liability to the community. This is changing because EPA wants to encourage the recycling of many components of HHW such as batteries and mercury-containing thermometers. EPA announced in 1995 that collection, storage, and transportation of some components of HHW by a voluntary program will not be subject to RCRA, the law applicable to hazardous waste. However, the waste will have to be either recycled or properly treated before disposal.

DISCUSSION QUESTIONS

1. (a) In the past, as you shopped for home products, how often did you check a product's label to identify its hazards?
   (b) Will your shopping habits change in the future to reflect the information you have obtained from this chapter? If so, how?

2. (a) Are there chemical products you would never have in your home? If so, provide examples and reasons.
   (b) What products would you only reluctantly have in your home?
   (c) Under what circumstances would you use pesticides your home or outside?

INDOOR AIR POLLUTION

In the late 1970s, EPA and Harvard University began the Total Exposure Assessment Methodology (TEAM) studies, which examined sources of pollutants and their transport and fate in the environment. Investigators monitored human exposure to specific pollutants by sex, age, smoking habits, and occupation. Investigators were startled to find that, regardless of the community examined – rural, lightly industrialized, or highly industrialized – indoor air pollution was the major source of human exposure to many air pollutants. Typical indoor pollutants are VOCs and several criteria air pollutants, including particulates. Pollutants have many sources, such as construction materials and numerous consumer products, such as cleansers, polishes, paints, moth balls, and glue. Pollutants are emitted from combustion appliances such as wood stoves, gas stoves, and kerosene heaters. They are also emitted by hobby materials, new rugs, furniture, drapes, rugs, and clothing. Chlorinated water releases them. Moist areas in the home encourage the growth of biological contaminants, which can become airborne. The gas radon seeps from rocks and soil under homes up into indoor air and also finds its way into indoor air from well water pumped into the home. Figure 13.1 displays some of the many sources of indoor air pollutants.

Exposure to indoor pollutants is especially significant because most people spend more than 90% of their time indoors. Ventilation is sometimes sacrificed to energy efficiency in newer homes. Older homes have their own

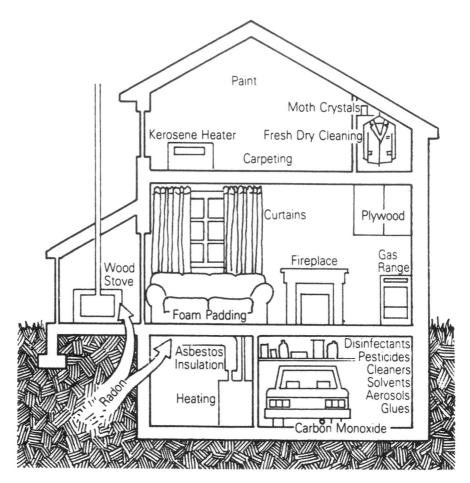

FIGURE 13.1. Sources of air pollution in the home. *Source*: U.S. EPA Office of Air and Radiation, Washington, D.C.

problems, such as flaking lead paint or more sources of biological contamination. Formaldehyde and nitrogen oxide are indoor air pollutants that can irritate eyes, nose, and throat. Carbon monoxide causes hundreds of deaths each year and, at lower concentrations, may cause flulike symptoms or headaches. Airborne microorganisms can cause allergies and infections. Radon is associated with lung cancer. In addition to acute effects of individual pollutants, chronic problems may arise from ongoing exposure.

General strategies to reduce indoor air pollution are source control and ventilation. Source control is eliminating or reducing pollutant sources; it is P^2. Good ventilation is a means to control pollutants that are produced. Exhaust fans should be vented to the outside. Especially in tight home, installation of an air-to-air heat exchanger to provide good ventilation with minimum heat loss can be useful. Electronic air cleaners and high efficiency particulate air (HEPA) filters clean air by trapping dust, fibers, pollens, skin flakes, and pet dander, but they do not ordinarily trap gases. To be effective, HEPA filters

Uranium-238
 ▽
 ▽
Radon-222
 ▽
Polonium-218 ▷ Lead-214 ▷ Bismuth-214 ▷ Polonium-214 ▷ Lead-210

FIGURE 13.2. Radon and its relatives.

must be changed regularly. Specific indoor air pollutants and strategies to reduce them are dealt with in the discussion that follows.

Radon

The ultimate source of ^{222}Rn (radon-222) is ^{238}U (uranium-238), a radioactive element naturally found in soil and rocks throughout the world. ^{238}U goes through a series of decay reactions to form ^{222}Rn (see Figure 13.2). If the radon – which is a gas – surfaces outside, it dissipates into the atmosphere. If it surfaces beneath a house or other structure, it seeps up into openings into indoor air. Because built structures are closed, radon concentration builds up in the air. Because radon is most likely to surface in the basement, concentrations are particularly high there. Radon in schools is a special concern. Trailers are spared because they are not in contact with the earth. Radon levels in outdoor air are typically between 0.1 and 0.4 pCi/L (pCi = picocurie, a unit of radioactivity), whereas the average radon level inside the home is about 1.3 pCi/L. The action level for radon in indoor air is 4 pCi/L; that is, if the concentration of radon is 4 pCi/L or higher, EPA recommends that action be taken to lower it. An estimated 6 to 8 million American homes have levels greater than 4 pCi/L. In addition to soil, well water is a radon source, especially water from deep wells. Municipal water, even if its source is groundwater, has lower levels because, as a gas, radon escapes into the air during processing. Once water is pumped into a home, some radon escapes during showering and other water use. However, only a small portion escapes from water into indoor air. Water containing 10,000 pCi/L releases only about 1 pCi/L.

Why is radon of concern? A radioactive substance undergoing decay emits ionizing radiation – an alpha particle, a beta particle, or gamma rays. For example, as radon decays to polonium-218, it emits an alpha particle. All three forms of radiation are capable of ionizing atoms by stripping electrons from them. In living tissue, this can result in damage to the genetic material, deoxyribonucleic acid (DNA). The *half-life* of a radioisotope indicates how long one-half of its atoms take to undergo decay. ^{222}Rn has a half-life of 3.8 days, so the average radon atom breathed into the lungs is breathed out again before it decays. But radon gas decays into solid radioactive daughter elements, which can attach to dust particles in the air. When the dust is inhaled, the solid daughter elements can be deposited onto airway walls and reside long enough to decay. ^{218}Po (polonium-218) and ^{214}Po (polonium-214), with half-lives of 3 minutes and less than 1 second, respectively, are especially likely to decay in the lung. The alpha particle they emit can damage the DNA in dividing lung cells. If the damage is not repaired, cancer can result.

Approximately 150,000 lung cancer deaths occur each year in the United States; an estimated 10% of these, or about 15,000, may be caused by radon exposure. Smokers and people in homes with smokers may have the highest risk, as a result of the particles released into the air during smoking. Radon daughters attached to the particles can be inhaled and trapped in the lungs. The situation is somewhat confused because a 1995 study indicated that non-smokers were at greater risk, leaving the issue as to who is at greatest risk un-resolved. Most risk arises from soil radon that has seeped into indoor air, not from radon that has escaped from water. This is because most of the radon in water, an estimated 99.99%, does not escape. High radon concentrations can build up in the air of a closed bathroom during showering, but the exposure time is short. People who drink water with high radon levels have a slightly greater cancer risk than those who do not.

How is radon detected and reduced?

- In air: Inexpensive test kits are available; in a typical short term test, a canister of activated charcoal is left open to the air for 2 to 7 days, then sealed and sent, as directed on the kit, to a laboratory for analysis. The laboratory sends the results to the homeowner. One high reading is not enough to justify immediate concern. The result needs to be confirmed through a longer-term test. Some states require testing for radon before a home is sold, but there is no federal requirement. A number of factors affect radon concentration in indoor air. Levels are lower in the summer, when a house is more open; thus, testing is ordinarily done in winter, when levels are highest. Various remediation strategies are available. Radon is sometimes much reduced simply by sealing the openings in basement floors. Or it can be diverted from a home by installing a pipe below the basement, which is connected to a vertical pipe that extends up the walls of the house and vents to the outside with the assistance of a suction fan. Remediation costs vary but average about $1,200. Those building a new home have the option to use P^2, that is, they can install a system to divert radon at the time the home is built. In locales with the greatest potential for high indoor radon levels, EPA encourages builders to install a passive radon control system at a cost of $350 to $500 (see Figure 13.3).

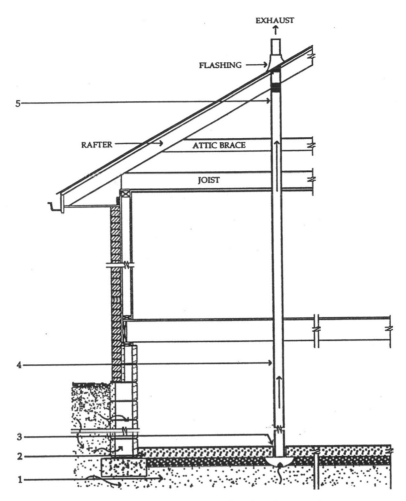

FIGURE 13.3. A passive system to control radon for new home construction. The system has five parts: (1) A layer of gas-permeable material such as gravel. (2) Over the material is a layer of polyethylene sheeting. (3) All openings in the foundation floor are sealed and caulked. (4) A polyvinyl chloride vent pipe that extends from the gas-permeable layer to the roof is installed. (5) Finally, provisions are made to install a fan if radon levels remain higher than desired. *Source:* U.S. EPA, *Pollution Prevention News*, March–May 1994, p. 5.

- In water: EPA has proposed a maximum contaminant level (MCL) for radon in water of 300 pCi/L; if it is adopted, municipal water supplies containing more than 300 pCi/L will have to reduce radon concentration. The northeastern states, which have water levels that are particularly high, would be most affected by high remediation costs. Recall that only an estimated one radon molecule in ten thousand escapes from water into a home's air when water is used. Consider water containing 300 pCi/L; 1 part in 10,000 of 300 pCi/L would increase indoor air levels by 0.03 pCi/L. If

BOX 13.1

Sources of ionizing radiation. As seen in Figure 13.4, more than 80% of the average person's exposure to ionizing radiation is from natural sources, and 55% of that natural exposure is from radon. The actual percentage for a particular person is higher or lower than 55%, depending upon the level of radon in that person's home. Another 11% of natural exposure is from radioactive elements within our own bodies such as ^{40}K (potassium-40) and ^{14}C (carbon-14). About 1 in every 10,000 potassium atoms in our bodies is radioactive. Terrestrial sources, rocks and soils, provide another 8% of natural exposure. ^{238}U is the largest terrestrial exposure, with smaller exposures from elements such as ^{232}Th (thorium-232) and ^{40}K. These radioactive elements, naturally found in rocks and soils, find their way into water, air, and food, and it is via these routes that humans exposure occurs. Another 8% of natural exposure is from cosmic rays, that is, the electrons, protons, and photons that enter earth's atmosphere from outer space. Cosmic rays react with elements in the earth's atmosphere to form ^{14}C and ^{3}H (hydrogen-3, or tritium). ^{14}C and ^{3}H settle to earth and find their way into plants and animals. ^{14}C dating is often used to determine the age of once-living plants and animals. About 300 cosmic rays per second pass through the body of a person living at sea level. Those living at higher altitudes have higher cosmic exposures, and aircraft crews and passengers experience higher still.

Even homes with only an average amount of radon generate interesting comparisons. Because of radon, the average nuclear plant worker goes home to higher levels of radioactivity than experienced at work. Radon came to national attention in 1984 because of an incident at a nuclear power plant in Pennsylvania. An engineer, Stanley Watras, set off the plant's radiation detector alarms. Investigation revealed that the source of the radioactivity was not the plant, but his home, which had a radon level of 2,700 pCi/L, 2,000 times greater than an average home. Materials taken from the earth, such as coal, also contain radioisotopes, and a portion escape into the atmosphere when coal is burned. The steam from geothermal energy also contains radioactive substances. Phosphate rock, used in fertilizers, contains higher levels of uranium than does surface soil, and groundwater contains higher levels of radioactivity than surface water.

Ionizing radiation produced by human activity represents about 18% of the average person's exposure. Most, about 83%, results from medical diagnostic procedures, especially X-rays used for chest and dental exams and mammograms. Most of the remaining 17% of anthropogenic radiation is from building materials and consumer products. Taken together, occupational exposure, nuclear fallout, and nuclear power represent less than 1% of exposure.

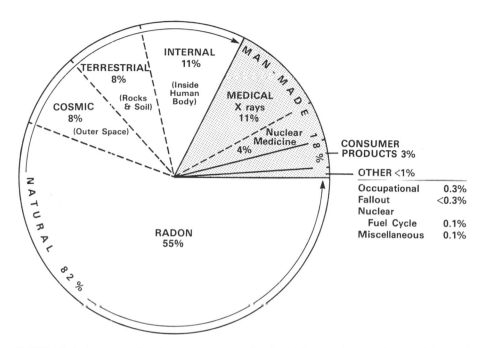

FIGURE 13.4. Sources of radiation exposure for the U.S. population. *Source*: National Council on Radiation Protection and Measurements, Bethesda, Maryland, Report No. 93, Figure 8.1, September 1987. Reproduced with permission.

the home already had an air level of 1.3 pCi/L, an additional 0.03 pCi/L from water would increase it only to 1.33 pCi/L, a small difference. But EPA has estimated that an additional 192 cancer deaths per year result from radon in water that has escaped into indoor air. This risk is greater than that for any other water contaminant that EPA regulates. There is concern that if radon is not regulated, industry can claim that toxic pollutants they release into water are much less risky than radon and therefore should not be regulated either. Others counter argue that society has the right to expect industrial emissions to be more strictly controlled than natural ones. EPA's science adviser suggested that the agency set a MCL not of 300 pCi/L, but of 1,500–2,000 pCi/L. An NRC panel requested that EPA make a new risk assessment of radon's health risks.

Scientific uncertainty about radon's risk. What could be less controversial than urging people to measure and, if necessary, to lower radon levels in their homes? A number of scientific groups, including an NAS panel, agreed that radon causes lung cancer in humans. Radon is unlike many other possible human carcinogens, for which only animal information is available. Nonetheless, there is disagreement about the risk of radon in air and water. The association between radon and human lung cancer was found by studying miners who worked in mines prior to the introduction of modern ventilation. These early mines contained radon levels hundreds or thousands of times higher than those found in most homes. Moreover, most miners smoked and the mines were extremely dusty. Some scientists doubt that lung cancer deaths among workers exposed to high levels of radon in very dusty mines can be extrapolated to the much lower exposures of average homes. A number of epidemiologic studies have examined possible relationships between home radon levels and lung cancer. A 1994 study reported in the *Journal of the National Cancer Institute* did not find a correlation between lung cancer and home radon levels. In this study 538 Missouri women with lung cancer who did not smoke were matched with women without lung cancer, and the radon levels in their homes were compared. Investigators concluded that the magnitude of the lung cancer risk from radon levels commonly found in United States dwellings appears low.

Some scientists argue that EPA's action level for radon, 4 pCi/L, is too low. They believe that EPA should concern itself only with homes containing at least 20 pCi/L. Only an estimated 50,000 American homes have such high levels. Critics also believe that EPA should inform people of the large degree of uncertainty as to whether radon poses a danger at levels found in most homes. EPA counters that there may be a risk from any level of exposure to ionizing radiation and that 99% of radon's calculated risk is to people exposed to less than 20 pCi/L. The agency also points out that although Canada has a 20-pCi/L action level, it is an exception. Other countries, Great Britain and Germany, for example, have action levels between 3 and 10 pCi/L.

BOX 13.2

Radon and lung cancer. In a study published in 1995, several National Cancer Institute investigators reanalyzed lung cancer deaths among uranium miners exposed to radon. They pooled data from 11 studies covering a total of 2,700 lung cancer deaths among 65,000 miners. They found that as the miners' exposure to radon increased, so did lung cancer risk. Their analysis suggested that even average home radon levels pose some risk. The study also found that long-term exposures to low doses, a condition similar to that found in homes, was more risky than short-term exposure to high doses of radiation. Although the authors understood the difficulty of extrapolating from conditions found in mines to those in homes, they recommended that homes with radon levels greater than EPA's action level of 4 pCi/L indeed be remediated.

DISCUSSION QUESTIONS

1. Assume that your home has a concentration of radon in air of 1.8 pCi/L, and that water piped into the home contains 750 pCi/L.
   (a) Assuming that 1/10,000 of the radon from the water is released into the air during use, how much will the home's air concentration of radon increase?
   (b) Do you believe that this increase is significant enough to justify removing radon from the water? Explain.
   (c) If not, what water concentration of radon would concern you enough that you would remove it?

2. (a) Before you buy a home, will you request that it first be tested for radon? Why?
   (b) If you already own a home, have you tested for radon?
   (c) What level would radon need to reach in your home for you to decide to remediate it?
   (d) If you had small children in the home, would you feel comfortable about their regularly playing in a basement that had not been tested for radon?
   (e) If the children only occasionally played there, would that affect your concern?

Combustion products

Environmental tobacco smoke is the most serious of the environmental health risks to which a person is exposed. In homes with a smoker, smoking is a major source of combustion products. Burning tobacco emits particulates and hundreds of chemicals, including carbon monoxide, benzene, formaldehyde, cadmium, lead, arsenic, even dioxins. One cigarette may not emit large amounts of pollutants, but people ordinarily smoke many cigarettes over the course of a day and, especially in enclosed surroundings, combustion products build up. In children living in a home with a smoker respiratory problems may develop or worsen. Smoking is also, of course, associated with development of lung cancer and of sometimes severe respiratory disorders, heart disease, and a number of other health problems. If a person does smoke indoors, there is a need to extinguish a cigarette completely after smoking it because a smoldering cigarette produces more carbon monoxide than one being actively smoked.

Not only smoking but any source of combustion in the home is a source of incomplete combustion products. Pollutants produced include the gases carbon monoxide and nitrogen oxides. Nitrogen oxides are most often found at high levels in homes with a gas stove or dryer or with a kerosene heater.

SOURCES OF CARBON MONOXIDE

1. Room heater
2. Furnace
3. Charcoal grill
4. Range
5. Water heater
6. Auto in closed garage
7. Fireplace

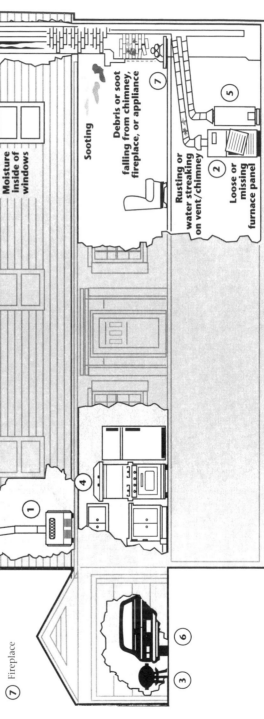

Loose or disconnected vent/chimney connections

Loose masonry on chimney

Moisture inside of windows

Sooting

Debris or soot falling from chimney, fireplace, or appliance

Rusting or water streaking on vent/chimney

Loose or missing furnace panel

FIGURE 13.5. Sources of and clues to a possible carbon monoxide problem. *Source:* U.S. Consumer Product Safety Commission, Washington, D.C.: U.S. Government Printing Office, 1993, O-356-764.

Efficient burning is ordinarily very desirable as a means to reduce the formation of incomplete products of combustion, but, as discussed in Chapter 5,
there is an exception to this rule: The temperature of a very hot flame promotes reaction between atmospheric nitrogen and oxygen to produce nitrogen oxides. These can irritate eyes, nose, and throat; can increase asthma
attacks; and may increase susceptibility to infection. Particulates, another
combustion pollutant, also contribute to respiratory and other problems.
PAHs, associated with soot, contribute to the problems caused by particulates.

Carbon monoxide in enclosed spaces caused 594 deaths in the United
States in 1991. At lower concentrations, this colorless odorless gas can lead to
headaches, dizziness, nausea, and other flulike symptoms. The Consumer
Product Safety Commission now recommends that carbon monoxide detectors be installed in homes. When concentrations of carbon monoxide reach
unsafe levels, an alarm sounds. Figure 13.5 indicates potential sources of carbon monoxide. To reduce the danger of carbon monoxide buildup in the
home, wood, gas, and coal stoves, furnaces of all types; fireplaces, and chim-
neys; and their connections must be regularly maintained. Although vigilance
is necessary for all furnaces, this is especially true of gas furnaces, which have
most often been the source of carbon monoxide poisoning. Kerosene heaters
can also be a major carbon monoxide source in the home and must be vented
to the outdoors. Other precautions against carbon monoxide buildup in the
home include installing appliances according to the manufacturer's instructions (ordinarily by professional installation), venting stove hoods to the outside, and not leaving a fire smoldering in a fireplace. If a fire is left burning,
close it off from the room.

Occupants in a home with a garage attached to or under it are exposed to
combustion products from the exhausts of automobiles, lawnmowers, and
other internal combustion engines. In addition to carbon monoxide, much of
the benzene in indoor air is from attached garages. Formaldehyde is also
found in car exhausts, andd stove and furnace emissions.

Building materials, home furnishings, and clothing

Many building materials, furnishings, and types of clothing emit VOCs. Formaldehyde is typically the one of most concern. It can irritate eyes, nose, and
throat and sometimes causes wheezing, coughing, fatigue, and skin rash.
Some people become sensitized to, that is, allergic to, formaldehyde. There
are many sources of formaldehyde, including resins and glues found in
pressed wood products such as particleboard, fiberboard, and hardwood ply-
wood; new furniture, drapes, and wall coverings; new carpets; and new permanent press clothing. New trailers often have many formaldehyde-emitting
sources, although regulations now require a seller to provide the buyer with

an information sheet on ventilation before a sale is completed. For especially sensitive individuals, formaldehyde home monitors are available. Other VOCs can also cause adverse reactions in sensitive individuals. The amount of formaldehyde emitted by many products has been reduced in recent years. Look for pressed wood products that have a United States Housing and Urban Development (HUD) emissions seal, because they emit smaller amounts of formaldehyde. Some wood products are now manufactured without formaldehyde. When possible, do home remodeling and install new furnishings in the summer, when the house can be well ventilated. Wash new permanent press clothing before wearing it. Have drapes cleaned before installing them or air them out in a place away from people until they lose their odor. Ask that new carpets be unrolled and well aired before installation in your home and, if adhesives are needed for the rug, ask for low-emitting ones.

Consumer products

Many consumer products contribute to indoor air pollution by emitting VOCs and, sometimes, particulates, including paints, stains, paint thinners and strippers, varnishes, and turpentine. VOCs produced by painting and related activities are the main indoor air pollutant; another source is pesticides, some of which are volatile. Other pesticides become volatile when discharged in aerosol sprays. Floor or furniture polishes and some waxes emit VOCs, as do the cleaners found in the bathroom and kitchen, and many personal care products such as nail polish, nail polish removers, and hair sprays. Motor vehicle products such as gasoline and oils emit VOCs. Sprays, especially aerosol sprays, contain tiny particulates and VOCs. Many art and craft materials are potentially hazardous, for example, paints, glues, and other materials give off VOCs. Other hobbies that generate indoor pollutants are model building, photography, ceramics, painting, and jewelry making. Woodworking generates significant amounts of particulates, as does pottery making. Metal working and stained-glass working produce metal fumes. *Rather than trying to remember many individual products, remember that almost any consumer product is potentially hazardous.* Read the label and follow instructions. If a product has the word *danger* or *poison* on the label, consider using less hazardous alternatives.

There are countless instances of acute adverse effects caused by improperly used consumer products. A woman became temporarily paralyzed after vigorously applying flea spray inside her home without adequate ventilation; a student fainted while using a spray cleaner on her bathtub in a small unventilated bathroom; another reported coughing and choking as a result of using an aerosol spray too near the face. Some people sneeze and tear up just from walking through a store aisle containing some of the products noted here.

Many reactions are not serious, but others can make a person miserable. Or, as in the case of the woman overexposed to the flea killer, the reaction may be life-threatening. Chronic effects are also possible, especially when a volatile product is used without good ventilation. Liver cirrhosis that developed in a Maine woman was attributed by her physician to paint stripper she had worked with over several winter months in a closed room. This is an extreme example, but no one is immune to adverse effects when simple precautions are not taken.

The solutions to consumer product emissions have been mentioned. In some cases, the product can be avoided. In most cases, good ventilation is critical to proper use. Use the minimum amount of product that will do the job. This is true not only of pesticides, which many people know are potentially dangerous, but also of products as "simple" as bathroom cleaners. Read labels. Artists and hobbyists can study information on practicing their craft safely; health and safety information is sometimes published by craft periodicals, and libraries often have craft books that include safety information. A prominent information source is the Center for Safety in the Arts, 5 Beekman St., New York, NY 10038, which will send a list of their publications on request. Many state health offices also provide safe guidelines for using art supplies, especially those for children.

Moisture

Uncontrolled sources of moisture generate mineral particles and microorganisms such as mold, bacteria, and viruses. Any of these can become airborne and be inhaled. Small particulates can be breathed deeply into the lungs and cause or aggravate respiratory irritation. Molds can aggravate allergies or give rise to infections. Sources of moisture include humidifiers, air conditioners, ventilation systems, and refrigerator drip pans. Other sources are damp attics and basements. When a cold wall exposed to moisture hits the dew point, moisture settles onto it and that wall can then serve as a growth area for mold. So can water-damaged areas in a home and water-damaged furnishings, moist papers, and books. Warm moist places are especially attractive to growth of microorganisms. Plants in the home can be attractive and healthful, but, kept moist, they too breed microorganisms.

A device sold at hardware stores can be used to measure indoor humidity. Recommended humidity levels are 30%-50% although some argue that anything higher than 30% encourages microbial growth. However, especially in summer, humidity may naturally be higher. If indoor air is too dry in the winter or in desert homes, some health specialists recommend drinking more water rather than installing a humidifier. If a home humidifier is used, use one that generates steam, not cold mist. If it generates cold mist, fill it with deionized rather than tap water. Deionized water does not contain minerals, which can become airborne particles. Regularly clean the humidifier to prevent growth of microorganisms. Properly maintain air conditioners, ventilation systems, and refrigerator drip pans by cleaning them regularly and changing air conditioner filters as recommended. Changing plant soils frequently is also recommended to minimize molds on wet soil.

Biological pollutants

Biological contaminants include more than microorganisms; they include pollens, animal dander, dust mites and their feces, cockroaches and their feces, cat saliva, and rodent urine. Microorganisms can cause not only infectious diseases, but other problems as well. Even dead airborne microbes can cause allergic reactions in sensitive individuals, as can bits of dead cockroaches or plants. About 15% of people are allergic to dust mite feces. Dust mite and cockroach exposure is associated with an increased likelihood of asthma. Dirty carpets often have dust mites, and they are also found on mattresses. Pollens are an example of contaminants that enter the home with outside air or are carried in on pets. Animal dander and cat saliva enter the home with pets. Especially in the South, cockroaches find their way inside, where their body parts and feces become part of a home's dust. After it dries, cat saliva or rodent urine can also become airborne. Depending upon the specific contaminants, many prevention and remediation strategies are possible. Maintaining a home so that cockroaches are not attracted and maintaining clean carpets are obvious steps. Consider wood flooring instead of carpeting, or use throw rugs, which can be washed. Pets may have to be banished from a home with allergic individuals. Other allergies require more detective work to determine the source and ways to remediate it. Libraries and bookstores are sources of more information on biological contaminants. A physician, especially an allergist, may provide literature or make recommendations.

Dust and dirt

Dust and dirt tracked in from outdoors may contain lead from old paint, pesticide residues from lawns and gardens, and many other contaminants. Once inside, these find their way into carpets and air. Carpets are almost impossible to keep completely clean. When carpets or wood floors are vacuumed or swept, large amounts of dust often become airborne. To lessen the amount of airborne dust during sweeping, lightly wet mop the floor beforehand. Plants, artificial flowers, and dried arrangements are truly "dust catchers."

DISCUSSION QUESTIONS

1. You are considering buying a home that is 35 years old and decide to check out its potential environmental hazards.
   (a) What materials and conditions in the interior and exterior of the home, basement, and attic will you examine?
   (b) What will you look for in the heating system?
   (c) What about the water system?
   (d) What about the home's outbuildings and land?
   (e) Repeat this exercise for buying a new or almost new house.
   (f) Repeat it for buying a trailer.

2. If you have a wood-burning stove or fireplace:
   (a) Is pollution from a wood-burning stove a great enough concern to you that you would buy a new air-tight stove? Explain.
   (b) Would your decision be different if you had children?
   (c) Do emissions from new carpets, drapes, or wood paneling concern you enough that you would install these only during the summer?
   (d) Would you apply paint or varnish to the inside of your home in the winter?

3. (a) What water pollutants may become airborne in a home served by municipal water?
   (b) What about a home served by well water?

FURTHER READING

American Chemical Society. 1987. *Pesticides, Information Pamphlet*. Washington, D. C.: ACS.

Cralley, L. V., Cralley, L. J., and Cooper, W. C. (eds.) 1990. *Health & Safety Beyond the Workplace*. New York: John Wiley and Sons.

EPA Journal. 1993. 19(4), EPA 175-N-93–027, Oct.-Nov..

Grossman, J. 1995. Dangers of Household Pesticides. *Environmental Health Perspectives*, 103(6), 550–4, June.

Harley, N. H. 1994. Radon Risk Revisited, *Health & Environment Digest*, 8(5), 1–3, Aug./Sept.

U.S.CPSC. 1993. *The Senseless Killer, Carbon Monoxide*, Washington, D. C.: CPSC.

U.S. CPSC, U.S. EPA, and the American Lung Association. *Combustion Appliances and Indoor Air Pollution*. Washington, D. C.: CPSC, EPA, ALA.

U.S. EPA. 1991. *Citizen's Guide to Pesticides*, 4th ed., Washington, D.C.: EPA, 22T-1002, Nov.

U.S. EPA and U.S. CPSC. 1988. *The Inside Story: A Guide to Indoor Air Quality*. Washington, D. C.: EPA/400/1–88/004, Sept.

U.S. EPA and U.S DHHS. 1992. *A Citizen's Guide to Radon*, 2nd ed. Washington, D.C. 20460: EPA 402-K92–001, May.

INDEX

Bold page numbers indicate a definition or description of the term.